# EARTH SCIENCE

# EARTH SCIENCE

## for Waldorf Schools

by
Hans-Ulrich Schmutz

translated by
Tom Wassmer

*Waldorf*
PUBLICATIONS

*Printed with support from the Waldorf Curriculum Fund*

Published by:

Waldorf Publications at the
Research Institute for Waldorf Education
38 Main Street
Chatham, NY 12037

Title: *Earth Science*
Author: Hans-Ulrich Schmutz
Translator: Thomas Wassmer
Editor: David Mitchell
Proofreader: Ann Erwin
Cover: David Mitchell
Cover image Bryce Canyon in Utah by David Mitchell

German Publisher: Freies Geistesleben
German ISBN: 3-7725-1687-4
Translated and printed with permission of the author

# Table of Contents

# Acknowledgements

The work presented here was entirely developed through teaching high school students; I owe them for stimulating impulses, onward leading questions, and great tenacity. The college of the Rudolf Steiner School Wetzikon (Switzerland) provided me with almost every freedom to design my classes, and with generous block time for earth science main lessons. The presentations became richer every year, and resulted in an increasingly clearer priority determination.

This project could be manifest thanks to the ideal and financial support of the Stiftung zur Förderung der Rudolf Steiner-Pädagogik in der Schweiz (Foundation for the Support of Rudolf Steiner Pedagogy in Switzerland) through G. Hug and R. Thomas. Cofinancing was provided by the Pedagogical Section at the Goetheanum, the Stiftung Evidenz ([Evidence Foundation] and the Pädagogische Forschungsstelle des Bundes der Freien Waldorfschulen, Abteilung Kassel [Department of Pedagogic Research at the Association of Free Waldorf Schools in Kassel].

I am thankful to the members of the workgroup "Curriculum Committee Geography" of the Pedagogic Research at the Association of Free Waldorf Schools in Kassel: I. Bechinger, M. Benner, E. Ch. Demisch, F. Kubier, M. v. Mackensen, K. Rohrbach, and G. Wolber.

For the critical revision of the manuscript I am especially thankful to R. Fried, M. v. Mackensen and R. Rist; and for the revision of individual chapters to I. Bechinger, R. Koehler, J. Kühl, M. Rist, K. Rohrbach, W. Schad, and R. Ziegler.

I also received correction proposals from students of the high school teacher training in Kassel and the Free University of Anthroposophic Pedagogy (Freie Hochschule für Anthroposophische Pädagogik) in Mannheim. G. Christen Terrani actively helped by making a fair copy of most chapters, and by providing and implementing many helpful layout suggestions.

Last but not least, I thank the translator and publishers for the careful preparation and the timely production of the book.

The translator wishes to thank the author, Hans-Ulrich Schmutz, for inspiring me as a teacher of earth science and for letting me translate his book. I want to also thank David Mitchell from AWSNA Publications for his support and patience with the localization of this book for the English-speaking world. Last but not least, I am very grateful to RSF Social Finance–Mid States Shared Gifting Group, and the Foundation for Rudolf Steiner Books for their financial support.

# Chapter 1

# Introduction to the Life and Earth Sciences in the Waldorf High School Curriculum

## The Origin of This Project

After I finished teaching at the Rudolf Steiner School "Zürcher Oberland" in Wetzikon [Switzerland] in 1997, I was asked to document the earth science blocks that were developed and implemented over eighteen years in composing and establishing the high school. I tried to do so in the present text. In order to allow the reader to better understand the block canon described below, the circumstances of the Wetzikon School shall be outlined briefly.

[In contrast to many other Waldorf/Steiner schools in Switzerland and Germany —tr.], the Rudolf Steiner School in Wetzikon offers one set of classes 1–12. [Many German and Swiss Waldorf/Steiner schools offer two parallel classes and by that two sets of classes 1–12 within the same school—tr.] Many students leave the school and directly enter into an industrial or professional training. For those graduates of the senior class in Wetzikon and other neighboring Rudolf Steiner schools who want to enter a college or university or need the "Matura" [high school diploma in Switzerland—tr.] for other professional goals, an autonomous "Maturitätsschule für Absolventen der Rudolf Steiner-Schulen"

[diploma school for graduates of Rudolf Steiner schools] was established, by which then an external "Matura" [Swiss federal diploma] can be achieved after 18 months of preparation. This way Wetzikon did not need to care about the "Matura" and other secondary school qualifications in designing the curriculum, especially since graduates could start the diploma school without entry exams. High school teachers only needed to make a non-binding recommendation.

Many of our high school faculty did not work full-time at the school and had other jobs outside of the school community. In addition it was desired that every colleague would spend some time on a [small] research project in a field of their choice. One of these research projects carried by several colleagues was the work on a curriculum concept in such a way that all blocks taught to a certain class grade would imbue and have a stimulating and inspiring influence on each other. In this endeavor, Rosemarie Rist was a strong tractive force who never tired.

## Indications for the Internal Shaping of Life and Earth Sciences Lessons during Classes 1–12

Figure 1.1 shows an outline of the sequence of life- and earth science blocks in keywords as they were, with a few exceptions, realized in Wetzikon. The foundations to do so were Rudolf Steiner's courses *The Foundations of Human Experience*, previously *The Study of Man* (GA 293), *Practical Advice to Teachers* (GA 294), and *Discussions with Teachers* (GA 295); *Conferences with Teachers*, 3 Volumes 1919–1924 (GA 300); as well as the *The Curriculum of the First Waldorf School*, *Rudolf Steiner's Curriculum for Waldorf Schools*, v. Heydebrand 1928, Stockmeyer 1951. However, most important was to observe the child and the growing-up youth and to evaluate how teaching gradually influenced the development of their spiritual capacities.

The course of the school days can at first be portrayed as a descent from the subconscious being-within the world of the spirit into the world of works and deeds (see Fig. 1.1]. Through developmental stages of a more awakened experience of the soulful and later the living, the student is introduced to the material world. This form impulse is continued into a new ascent. This ascent needs to be stimulated during high school by tackling the developmental stages of forming judgments and opinions within the inanimate world most carefully. The student first exercises forming judgments and opinions related to the inanimate, inorganic world. Then he moves on to an understanding of the animate living. Through the intentional soulful / animate, he works towards the reflection of the essential question of sense and meaning by which he will gradually grow into a more conscious connection with the world of ideas and concepts. He has to move away from the naïve being-within the world

of context in order to gain the option to reconnect consciously in free initiative with the world of ideas as a grown-up.

### Life Science

Life science as an independent subject block starts in the 4th grade. Up to then, the student experiences the relation of the human being to the various kingdoms of nature, first in the form of fairy tales, then as plant legends and animal fables. Apart from the Creation story, the 3rd grade focuses on the agriculture and house building blocks. The student experiences how the human being shapes the world around him and his fellow beings using archaic crafts and trades. Animals as living-animate beings are the focus of the fourth school year. The animal science [zoology] block is introduced by elementary human science [anthropology] describing the threefold nature of the human being. The 5th grade pays tribute mainly to the processes of life. Due to its double meaning, the plant science [botany] block has its important place through the years. On the one side, the sequence from the lower to the higher plants within nature is reminiscent of the development of the childlike soul, whereas the contemplation of the plant cover from pole to Equator provides an imagination of the entire living earth. In the 6th grade, life science and earth science interconnect in a more narrow sense. A rock and minerals science [geology] block studies the earth which became solid and crystalline.

This walk through the realm of the human being and the three realms of nature has a well defined position and clear sequence within the middle of the first eight school years. The two last years of middle school and the 9th grade are reserved for human science [anthropology]. During the 7th grade, shortly before students finally lose the subconsciously experienced connection with the spirit-interwoven being of the realms of nature, the human science block discusses the relations between the physical human being and earth. On the face of it, the students learn a lot of useful facts for life about food, clothing, hygiene and housing. Placed at the right moment, this block serves to cover questions of nutrition—not seen from the point of view: "What is good for me?" but, according to Steiner, as a matter of making use of the last subconscious connection between the juvenile human being and the forces of nature to experience the harmony between the environment and the human organism and as a result effectively overcome selfishness. The human science block of the last middle school class deals with the human bone structure entirely in a mechanistic-physical point of view.

The topic of the skeleton is taken up again in 9th grade but it is now related to tendons, ligaments and muscles. The opportunity for active and spontaneous movement in the human being is investigated. The 10th grade contains the central movement blocks. In human science it is the internal living movement processes already presented in the 7th grade that are studied in the processes of the internal organs. On the other hand, the kingdom of minerals is taken up again, by making place for a block of crystallography (cf. chapter 4). After a long break, there is another direct study of plants in the 11th grade. It is done in such a way that cell science [cytology] leads into botany. In the years between, the students are regularly involved with plants during horticulture / gardening classes. The final year is especially dedicated to animals, which are investigated with regards to their *bauplan* [structural plan] as well as in terms of the sequence of their temporal appearance on earth (see chapter 9). The human science block of the

11th grade can be dedicated to the human biography while the seniors are ready for the embryology block, this block followed by a study of anthropology up to the beginning of classic history (see chapter 9).

### Earth Science [Geography]

Earth science also gets its first subject block during the fourth school year. It is home science (local history, geography, and natural history] during which the familiar surrounding area of the class is designed as a sophisticated map. The student adds all disconnected parts that are perceived through isolated senses and understood by the mind into the form of a map image and by this into a higher conceptual context. Here Steiner's curriculum indication means the training of causal concept formation, namely the experience of the context of soil, cultivation, climate and history. It is therefore not only about the formation of imaginations but about the fundamental grasping of spiritual connections.

On one side, the 5th grade geography lesson is embedded into plant science [botany], looking at the differences in the plant cover around the entire globe. On the other side, the actual geography block of the 5th grade starts from the more narrow understanding of home that was introduced in the 4th grade and leads to neighboring areas by "walking" around, e.g., along the course of rivers. A first country science [regional geography] is taught.

The 6th grade is much influenced by a rock and mineral science [geology] block enriched by many geographical concepts, especially those encountered during a working week away from school [geology field trip]. It is, on the other side, the privilege of the geography block, to develop the European [and American] landscapes

# Thinking about sense and meaning

to the sphere of the prototype-like **Spiritual**

in the sphere of the intentional **Soulful**

to the sphere of the animate **Living**

in the sphere of the **Inorganic-sensational World of Works and Deeds**

**Ascent**

Improvement in practicing judgment

During high school thinking serves the formation of the conscious soul (Bewusstseinsseele); Grasping the non-percetable.

---

*Zoology Bauplans (structural plans) of animals*
HS: *Embryology*
**A: Paleontology and Anthropology:** Evolution (development) of life and the human along the changing earth
**B: World economy:** global actions of humans

*Botany, taking cytology as starting point*
HS: *The human biography: from birth until old age*
**A: Astronomy:** Interactions between Earth and celestial bodies
**B: Energy Management:** Energy conversion, resources and alternative energy

HS: *Metabolism of the internal organs*
**A₁: Crystallography:** geometrical thinking
**A₂: The Earth in motion** in the air-, water- and rock cover
**B: Textile technology:** Interplay between landscape and inventions

HS: *The skeleton in motion: dynamics through tendons, ligaments and muscles*
**A: Geology:** the mountain backbone (Gebirgskreuz) is moved (earthquakes, volcanism, mid oceanic ridge, orogenesis)
**B: Function of modern life:** Tours of factories or plants, e.g. technology of waste processing

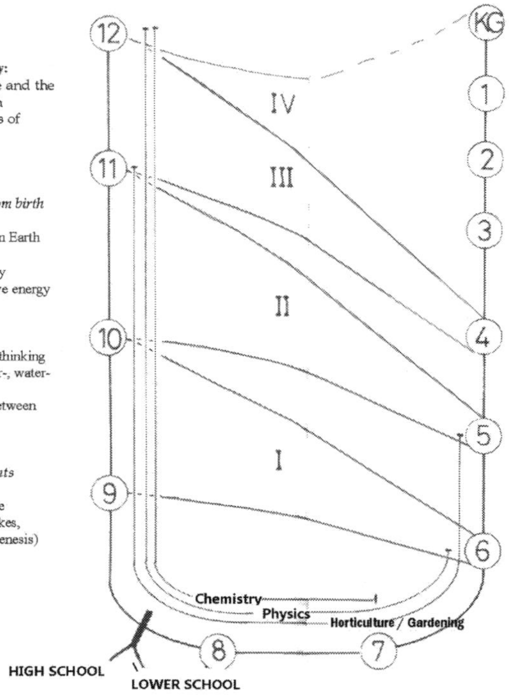

**HIGH SCHOOL**

**LOWER SCHOOL**

HS: *The human skeleton Mechanistic-physical*
**A: Mountain backbone (Gebirgskreuz):** the static, morphologic skeleton of the earth
**B: Geography of continents:** e.g. polarity of continents: "Old World" - "New World"

Chemistry / Physics / Horticulture / Gardening

HS: *The human within the world (Feeding, hygiene, clothing, housing)*
**A: Astronomy:** the universe (cosmos) around the earth
**B: Geography of continents:** Continuation Geography of continents: enter new areas, e.g. Africa (more focus on cultural issues)

9  10  11  12  8  7  6  5  4  3  2  1  KG

*Human three-foldness and animal science (zoology) epoch*
**A:** home science (local history, geography, and natural history)

*Soul development of the human as plant science (botany) epoch*
**A:** Plants as life of the earth from the pole to the equator
**B: Regional studies:** Wandering from home into new areas

*Imagination, forgetfulness and memory of the human as rocks- and minerals epoch*
**A:** Rocks and minerals (local history,
**B: European (American) geography:** new landscapes, developed from the underground (rocks, rivers, transportation)

*Creation story/farming - house building - crafts epoch: practical work, actions of the human in the world*

*Fairy tails*

*Plant legends animal fables*

*Play, everything is alive, soulful and spirited (inspired)*

---

# Experience sensational concepts

**Descent**

the attachment to the physical-alive soulful-**Spiritual** diminishes

Awakening on the world of the **Soulful (Beseelten)**

Awakening on the world of the **Living (Lebendigen)**

Awakening on the world of the Rigid, **World of works and deeds**

---

**Italic = indications for life science. HS = Human Science (Anthropology). A = General Earth Science epoch. B = Geographic epochs (regional geography, economy, also technology, life skills). KG = Kindergarten.**

I = 1. Kingdom of nature: grown inherent —»World of works and deeds (physical-mineral)

II = 2. Kingdom of nature: deal metamorphic with the surrounding conditions — »World of Life (Etheric)"

III = 3. Kingdom of nature: intentional shaping of processes — "Soul World (Astralic)"

IV = 4. Kingdom of the human: individual shaping the spirit —»World of Ideas (Prototypes, I-like)"

In the lower school thinking is in the service of the formation of the reasoning soul (Verstandesseele); Knowledge is gained, formation of imaginations

Fig. 1.1: Diagram of the internal shaping of the life and earth science classes during the classes 1-12; example: Rudolf Steiner School "Zürcher Oberland" in Wetzikon (Switzerland).

*Fig. 1*

from the geological underground and the river network.

The 7th grade student is moving along two tracks into the unknown: First, a fundamental astronomy block is indicated, during which the course of the celestial bodies is observed from a point of view on earth. The other unknown is a new continent, e.g., Africa or Asia. The points of contact with the history lessons are obvious as they cover the experiences of the conquerors and seafarers [sailors] at the beginning of the modern times. As the geography lessons of earlier grades focused more on landscapes and the conditions of travelling and trading, it should now focus more on the cultural conditions.

In analogy to the human skeleton, the final class of the middle school focuses on the mountain ranges throughout the world. *The Mountain Backbone of the Earth* (Steiner 1924a] is first to be investigated according to a morphological point of view. In this context, the geography of the North and South Americas can be covered together with Antarctica.

The earth science blocks of the high school classes are the topic of the present book. A chapter on the science of knowledge is inserted to allow a better understanding of the book's design.

## Stages of Judgment Formation: From the Reflection to the Archetype

In high school, in which the realms of nature occur in an order reversed to the areas of experience in the lower school, it is possible to get the training of judgment formation of world phenomena into a sequence of steps. First, students continue to practice judging the lifeless / inanimate world in order to step to thinking through the living / the alive in 10th grade. In 11th grade, they practice judgment in the sphere of the soul, and in their senior year, they feel their way to get a grasp of the ideal-spiritual. The formation of judgment is shown in an everyday example in order to allow a better soulful observation of this sequence of steps: "The entity/phenomenon in front of my room's window is a birch." How much of what needs to happen so that I can arrive at this accurate judgment? First, an experience of the necessity of a line-up/ separation can be observed, for observation is standing on the opposite side. I am here, and outside of me / separate from me / opposite to me is an uncertain something. Out of this attentiveness, the crucial thing happens: On the one side, opposite of me—conveyed by various sensory organs—stands the perceivable in countless, disconnected details. On the other side, I have dissolved out a concept from the universe of the world of concepts / ideas. This is because concepts not only have the quality to link together according to their content, but they are also able to point beyond themselves to the perceivable [the non-evident]. As a result of this, the definite, observing, directional relationship [thinking eye] towards the perceivable is being created. It is only due to this fact that the perceivable becomes observable. Therefore, the "thinking eye" is the instrument of focusing attention towards the perceivable in general.

In *The Science of Knowing: Outline of an Epistemology Implicit in the Goethean Worldview* GA 2, Steiner describes these facts in the embryonic condition. In the chapter "An Indication as to the Content of Experience" he writes:

> Let us now take a look at pure experience. What does it contain, as it sweeps across our consciousness, without our working upon it in thinking? It is mere juxtaposition

in space and succession in time, an aggregate of utterly disconnected particulars. None of the objects that come and go there has anything to do with any other. At this stage, the facts that we perceive, that we experience inwardly, are of no consequence to each other.

– Steiner 1886, chapters 1 and 24

And further:

At this level of contemplation, the world is a completely smooth surface for us with respect to thought. No part of this surface rises above another; none manifests any kind of conceptual difference from another. It is only when the spark of thought strikes into this surface that heights and depths appear, that one thing appears to stand out more or less than another, that everything takes form in a definite way that threads weave from one configuration to another, that everything becomes a harmony complete within itself.

– Steiner 1886, 4, 25

In the chapter "Thinking and Perception," Steiner elaborates on the process of thinking:

We would have to renounce our thinking entirely if we wanted to keep to pure experience. One disparages thinking if one takes away from it the possibility of perceiving in itself entities inaccessible to the senses. In addition to sense qualities there must be yet another factor within reality that is grasped by thinking. Thinking is an organ of the human being that is called upon to observe something higher than what the senses offer. The side of reality accessible to thinking is one about which a mere sense being would never experience anything. Thinking is not there to rehash the sense-perceptible but rather to penetrate what is hidden to the senses. Sense perception provides only one side of reality. The other side is a thinking apprehension of the world. Now thinking confronts us at first as something altogether foreign to perception. The perception forces itself in upon us from outside; thinking works itself up out of our inner being. The content of this thinking appears to us as an organism inwardly complete in itself; everything is in strictest interconnection. The individual parts of the thought-system determine each other; every single concept ultimately has its roots in the wholeness of our edifice of thoughts. (Ibid. 3, 48]

The connection between the spheres of the world of concepts / ideas and the world of perceptions that seem at first apparently separate, happens as follows:

In all cognitive treatment of reality the process is as follows. We approach [literally, gegenübertreten: step/move on the opposite side of—tr.] the concrete perception. It stands before us as a riddle. Within us, the urge makes itself felt to investigate the actual what, the essential being, of the perception, which this perception itself does not express. This urge is nothing else other than a concept working its way up out of the darkness of our consciousness. The mute perception suddenly speaks a language comprehensible to us; we

recognize that the concept we have grasped is what we sought as the essential being of the perception.
– Steiner 1886, 6, 49f

Let's go back once more to our gesture of thinking. Internally feeling all over I have to do the examination and decide: Is the entity on the other side dead or alive? Is it part of the realm of plants or the animal kingdom? I therefore have to decide, which of the universal concepts—lifeless, alive, technical, mineral, vegetable, animal—applies to the perception complex. After realizing that the concept "plant"—which, as previously mentioned, I must have grasped in an ideal sense—applies to the perceived, it is a matter of specifying and differentiating the general plant being. Here the transition takes place from the phase of intentionalization towards the direction of the universality of the perceivable to the phase of metamorphosing. The intentionalization still happened within the universality. The general concept "plant" is in contrast already referring to something perceivable, which had led to the archetype "plant." Now the sub-universality "birch"—which is also a general concept—has to be led towards the perception that already allowed the formation of the judgment "plant." The metamorphosing from the archetype "birch" takes place, which points beyond itself to perceivable entities [intentionalized] and then transforms itself to be similar to the concrete forms of these specific birches and then is finally inhered. Steiner describes this event in chapter 16, "Organic Nature:"

Following Goethe's example, let us call this general organism *typus*. Whatever the word typus might mean etymologically, we are using it in this Goethean sense and never

mean anything else by it other than what we have indicated. This typus is not developed in all its completeness in any single organism. Only our thinking, in accordance with reason, is able to take possession of it, by drawing it forth, as a general image, from phenomena. The typus is therewith the idea of the organism: the animalness in the animal, the general plant in the specific one. One should not picture this typus as anything rigid. It has nothing at all to do with what Agassiz, Darwin's most significant opponent, called an "incarnate creative thought of God's." The typus is something altogether fluid, from which all the particular species and genera, which one can regard as subtypes or specialized types, can be derived. The typus does not preclude the theory of evolution [Darwin called his theory descent with modification—tr.]. It does not contradict the fact that organic forms evolve out of one another. It is only reason's protest against the view that organic development consists purely in sequential, factual [sense-perceptible] forms. It is what underlies this whole development. It is what establishes the interconnection in all this endless manifoldness. It is the inner aspect of what we experience as the outer forms of living things.

The Darwinian theory presupposes the typus. The typus is the true archetypal organism; according to how it specializes ideally, it is either archetypal plant or archetypal animal. It cannot be any one, sense-perceptibly real living being. What Haeckel or other naturalists regard as the archetypal form is already a particular shape; it is, in fact, the simplest shape of the typus. The fact that in time the typus arises in its simplest form first does not require that the forms arising later be the result of those preceding them in time. All forms result as a consequence of the typus; the first as well as the last are manifestations of it. We must take it as the basis of a true organic science and not simply undertake to derive the individual animal and plant species out of one another. The typus runs like a red thread through all the developmental stages of the organic world. We must hold onto it and then travel with it through this great realm of many forms. Then this realm will become understandable to us. Otherwise it falls apart for us, just as the rest of the world of experience does, into an unconnected mass of particulars. In fact, even when we believe that we are leading what is later, more complicated, more compound, back to a previous simpler form and that in the latter we have something original, even then we are deceiving ourselves, for we have only derived a specific form from a specific form.

– Steiner 1886, 15–17, 78f

Let us return to our example: I therefore come from the general idea of a plant to the lignified, perennial tree, then further to a deciduous tree, and finally to a birch-like tree. The being birch is a sub-concept of the typus tree. During the last step—the step of inherence formation—I have to judge, whether the conceptual-evident parts of the observation process freely and seamlessly connect to the perceptual-given by forming a specific shape—here: to this specific birch in front of my room's window. The concept is now grasped, withheld from the perceivable, and therefore becomes a precise imagination, a perception judgment: It is a birch!

From this course of events it becomes obvious, that external circumstances such as climate, surroundings and time only support or inhibit conditions and are never the cause of the arrangement of the typus to become the actual birch. A precise portrayal of this complex course of events in terms of the science of knowledge is presented in Herbert Witzenmann's 1983 *Structural Phenomenology*. Therein it is elaborated, how the world can be understood as a magnificent morphogenesis, coming from a universal idea and guided through a sequence of steps called actualization, intentionalization and the metamorphic adaption to an individualized form. According to this, a concept can be a) actualized, which means dissolving it from the realm of universalities, thinking it as a singled-out concept in evidential activity; b) intentionalized, by using its many and diverse linkings, for example, towards the direction of the perceivable; c) metamorphosized, which means, to adjust it to the perceivable, and d) inherized. The concept can therefore be defined for a specific form, which means it is individualized, and the particular, the perceivable is embedded within the overall context.

## Sequence of Steps in the High School Earth Science/Geography Blocks in Europe

Let us now set out the sequence of steps for the high school classes. When describing the path from the archetype [idea] to the reflection [mental image] as we did before, we discover that in high school lessons the student walks the opposite path from the reflection to the archetype by learning by experience to move within these different levels—if they and the teacher are aware of them—without knowing their foundation in terms of the science of knowledge.

The geology block in 9th grade (cf. chapter 2) has the development of the student's logical thinking as its goal, starting with the rocks that have fallen out of the processual events to think their way into the process of rock formation. The student learns to direct his eye on the inhering of concepts: It is his task to assimilate the solidified configuration, which was released from the process of life and retained life only as a reflection, and to take it into a larger context, which includes the Becoming of this formation. However, this Becoming takes a causal course, as perceivable causes have to be found for each and every change. For example, the uplift of the Alps can be on the one side objectively based on the collision of the European and the African continents or on the other side on the lowering of the debris plains of the foothills north and south of the Alps.

In the following school year, the focus lies on practicing the metamorphic adaptation of a concept on a consolidated entity. A block on crystallography (cf. chapter 4) is a good option for practicing the transformability of a concept [here the tetrahedral principle] step by step in a controlled fashion. The second earth science block, "Earth in Motion" (cf. chapter 3), is as a kinematic discipline concerned with the correlation of assembly and disassembly within the air (atmosphere], water- (hydrosphere) and rock-spheres (lithosphere) of the earth. The basic principles of the activity of an organism are apprehended in thinking, i.e., the process of adaptation of the idea of an organism to the conditions of the surroundings are the focus of a question-answer search process.

The layered atmosphere of the earth may be used as an example. The ratio between transmittance and delimitation of sunlight, ultraviolet radiation or infrared radiation leads to a relative stabilization of the global temperature and by this to a stabilization of a climate suitable for life. When now approaching the nature of the task in the 11th grade, it is obvious that it can be tackled only with a look towards the aim of the 12th grade: on the activity of an internal visit of the causing entity, on the thinking grasping of the motives of development / evolution [Entwicklung], thus on the search for the archetype.

The astronomy block in grade 11 (cf. chapter 7) attempts to approach the soulful-intentional correlations between the earth, the sun, the moon, the planets and the stars. Cosmic rhythms are related to earthly rhythms, which can be found in the most diverse ways within the life processes on earth. First steps are therefore taken, to approach the morphogenetic forces that cause the transformation of the shape of the earth and which must be thought to be ideal.

In 12th grade, the relationship between the cosmos and the earth on the one side and the human being on the other side should be the focus of a concluding study (cf. chapter 9). It is set forth, in paleontology and likewise in anthropology, how life on earth unfolded from a universal typus in a step-by-step differentiation. The context of [the meaning of] life from the past into the future is made a subject of discussion. Lessons in economic geography which are more oriented towards life science and technology (cf. chapters 5, 6, 8 and 10) are arranged in an identical sequence to the steps of making a judgment.

The human science block of the 7th grade is taken up again in a more differentiated way. In 10th grade, the step from craft professions to the modern industrial society is done on the basis of textile technology. In a next step, another cornerstone of human civilization is investigated, namely energy conversion. This block lends global ecology a certain air. The 12th grade reserves the right to become familiar with the complex structures of the world economy that becomes more and more complicated by looking at the allocation of essential foodstuffs.

Let's go back once more to the starting point of our examination. There it was stated how the human being as a realizing entity connects concept and perception in a fourfold sequence of steps. The perceptions are related to a world which is at first inscrutable. On the assumption that the world evolved from universal, ideal archetypes, the cognizant human being has to go the opposite way. He has to regain the conscious approach to the archetypes on the path of cognition and by observing the world. One can illuminate this practicing path with an example in the activity of the high school student during earth science lessons (Schmutz 1993]: The 9th grader gets familiar with what the world has become by studying the "physical body" of the earth, which means that he practices applying concepts appropriately, for example to relate laws of nature to rigid forms as reflections of an earlier effectiveness. The 10th grader studies the vital life processes of the entire planet earth; he gets to know the organological principle, according to which transformation of the shape leads to morphogenesis. He gets to know the "etheric body" of the earth. He practices the metamorphosing of a concept illustrated by crystallography. The 11th grader turns to the cosmos and experiences how earth and cosmos stand by one another and form an intrinsic, sensible relationship. He practices how concepts can be related to the perceivable

that is subject to the transformation of the shape [metamorphosis]. Thus, he ventures into practicing the optional intentionalization of a general concept. The 12th grader experiences how in the evidence [actualized concept]—that is to say, by connecting the subjective act of thinking, which is determined back through the content of thoughts—an archetypal entity becomes, through the sequence of steps of intentionality and metamorphability, the reflection in form of what has become the perceivable world. Is the process of thinking the "I or ego" of the earth? By doing this the adolescent experiences him- or herself anticipatively as a free being, who is able to walk on the mental path from the archetype to the reflection and back to the archetype. This way he becomes a creator of the future. It should be added, that these four steps are always interconnected, which results in an unconscious or conscious, lightning-fast to and fro between these steps.

During each high school lesson, one element of the activity of thinking is given priority in practicing. My intentions are successful if these rough explanations make it possible to clearly show that the curriculum, developed by Steiner, is the work of a genius, presenting as much a study of the self of man as it is a study of the nature of the earth. The fundamental structure of every activity of [spiritual] knowledge brought the world into manifestation/appearance and reappears within the human being when he grasps and comprehends the world, the act of [re]cognition. Just as the human being himself, the Earth can be viewed as a being, which belongs to the world of universal ideas. Following this insight into the nature of things, the possibility arises to act in the world in a meaningful way by continuously gaining more comprehensive insights.

Indications from the study of man about the individual age groups are embedded within the following descriptions of the main lesson blocks.

## About the Characteristics of the Block Reports

The following block descriptions should always be assessed under the condition that the students go through all of these blocks. This assumption was always my starting point when designing my lessons and defining priorities. Therefore, each block should be seen in the context of the prior and subsequent blocks. The benefits from this reading will be the more, as the teacher turns his gaze on the year-by-year stepwise changing coverage of the subject. You may make the accusation that the choice of subject is one-sided, especially in grades 11 and 12. Whoever is more interested in the geography of economics and ethnological [anthropological] block descriptions should consult the book *The Living Nature of the Earth* (Göpfert 1999).

Rock formations, Wyoming. Photo by DSM

# Chapter 2

# Geology—Ninth Grade

## Introductory Observations

### Grade 9—The Transition between Middle and High School

In the first part of this chapter, we will describe the situation of the 9th grader according to the study of man [human science, anthropology]. On the one side, grade 9 stands at the beginning of the four years of high school—on the other side, at the end of the lower school phase, in which the gaining of knowledge and abilities depends strongly on the authority of the teacher.

In his lecture "Education for Adolescents," [GA 302a from 6/21/1922], Steiner brings out very clearly how a complete change in the lessons has to happen in an abrupt manner during the transition between grades 9 and 10 from acquiring knowledge to practicing the formation of judgments (Steiner 1922a). Until now questions about the "what" of a thing or an event were prominent in a more pictorial teaching style. The question "why" should be the focus of teaching from grade 10 on. Steiner points out that especially causal connections in the sphere of the inorganic should shine through as a formation of judgment starting at the 12th year of age [e.g., during the 6th grade physics block]. However, the main focus is still on the acquaintance of physical facts.

With this in mind, the nature of the task for grade 9 can be expressed like this: Besides a comprehensive getting to know of still unknown world appearances, above all, causal connections should light up within the soul of the young person, so the students can be increasingly won over with the matter itself and not as strongly arranged by the teacher being a raw model. In this context, the change from having just one class [grade] teacher to a multitude of subject teachers is a great help—as long as the subject teachers are not teaching in the way of thinking of class teachers. It is therefore not the change from the above mentioned "what" to the question "why" that is of immediate importance, but the student's emancipation from the authority of his/her teacher and onto his/her path to an independent way of working.

The world's appearances/phenomena should be preferably presented in polarities. For the students on the threshold of a conscious formation of knowledge, this lining-up/comparison, achieved from the phenomena themselves, falls on fertile ground. When looking at what happens during instruction in grade 9, the motif/ theme of polarity appears in many instances. In art instruction, painting with colors is replaced by black-and-white drawing. The German block strives from the polarity of the founders of the classical period of German literature Goethe and Schiller and in an English block of the transcendentalists in American literature. A second English block will focus on the polarity of comedy and tragedy. The human science block includes a review of the organization of muscles, especially the polar interplay between flexors and extensors. In physics, machines and equipment such as the telephone are reviewed by imagining various polarities with the thinking eye: compression/dilatation, on/off, reception/playback.

One peculiarity of causality within the realm of the inorganic world is especially important for instruction in grade 9. At the beginning of the fourth chapter of *Goethean Science,* GA 1, Steiner writes very clearly about the appearance of the study of inorganic nature:

> If it is a matter then of comprehending such a phenomenon [of the inorganic nature], this can be achieved only by our transforming into concepts what is directly there for the senses. We would succeed in this to the extent that nothing of a sense-perceptibly real nature remained that we had not permeated conceptually. … That which offers itself to our senses must appear as a necessary consequence of what we have to postulate ideally beforehand. If this is the case, then we can say that concept and phenomenon coincide. There is nothing in the concept that is not also in the phenomenon, and nothing in the phenomenon that is not also in the concept. … The important fact arises here that the sense-perceptible processes of inorganic nature are determined by factors that likewise belong to the sense world.… A conceptual grasp of such processes is therefore nothing other than a tracing of something sense-perceptibly real back to something sense-perceptibly real.
>
> —Steiner 1884, 4, 70f

This makes the direct connection between concept and appearance perfectly clear, as outlined in *The Science of Knowing: Outline of an Epistemology Implicit in the Goethean Worldview* (Steiner 1886]: the connection between experience and thinking. In the area of understanding the inorganic world, the act of thinking is aroused by the sensual appearances/phenomena. It leads to a preliminary closure, without the need to go beyond the realm of the sense-perceptible/tangible objects. As the cause of an effect belongs to the realm of the perceivable, the comprehension of the inorganic world is still able to support itself against the sense-perceptible phenomena. This graphicness is especially helpful for the 9th grader. A lower or middle school student gains knowledge by observing and naming concrete, objective phenomena. In terms of the process of thinking, the close relationship between the perceivable and the conceptual is experienced in a un- or preconscious manner. Happily he gets to know the phenomena of the world. This characterizes the analytic procedure. The student strengthens his/her intellectual faculties and gains confidence in this activity. An intensification of this process is the conceptual grasping of inorganic nature phenomena in form of the slightly more awake intellectual comprehension of causal connections. The 9th grader is still able to rely on the sense-perceptible when making connections. However, he already partially wakes up in a joyful experience when he/she forms causal judgments. This means walking along the path that leads to synthetic judgment.

This way it may become understandable that the inner characteristic style of teaching in grade 9 is still closer to that of middle school [grades 6 to 8] instruction. Confidence in the activity of thinking, which is more detached from the sense-perceptible, will not be achieved before grade 10. In summary, it can be stated that the aim of grade 9 is to take what has become the solidified image of the world as a starting point and, by asking questions and by thinking, move into the processes that created these images. The tethers for this process are the natural laws of the inorganic world, which were grasped experientially during the lower grades years. This leads to confidence in dealing with causality. This confidence is needed for success in the next step: when dynamic entities need to be intellectually grasped and detached from the sense-perceptible causes.

### Why Geology as the Earth Science Block in Grade 9?

One aspect in the earth's process of becoming is solidification and is thus mainly subject to inorganic natural laws: the rocks and minerals, which form the foundation for the turbulent and living/animated envelopes/sheaths of the earth. These lifeless images/reflections of the many and diverse processes of becoming should be the starting point for the study of the earth. If one wants to understand rocks and the connection between various types of rocks, he has to raise and look into questions of the formation and "back-formation" of rocks. In realizing how especially the events of earthquakes and volcanism, of the formation and back-formation of mountains, are strongly set in polarities, one realizes how rewarding it is, especially for 9th graders, to study Geology.

Note has been taken repeatedly that an examination of the underlying processes of motion—that is, plate tectonics—would be too early and would not fall onto fertile ground before grade 10.

### Recourses to the Middle Grades

The block described below is based on the assumption that, in grade 6, students studied rock types and their corresponding landscapes for several weeks—both in the classroom as well as outdoors (comp. chapter 1). As far as possible, concepts about the formation of granite, limestone, shale, sandstone, gneiss and marble were presented in a lively manner. Subsequently, various observation exercises were conducted outdoors. Spatial imagination was trained in various ways during geometry and drawing lessons. Additional revisits to physics and chemistry lessons can be made, wherein students learned about mechanical, hydraulic, and thermodynamic phenomena. While the 8th grade geography block dealt with the morphologic mountain backbone of the earth, now in grade 9 it can be understood in terms of process. All of these elements of knowledge and capabilities are now being recalled and applied with tact and care, as it is now up to the subject teacher to arrange the recourse, which relates to the time and realm of the grade teacher.

### Keeping the Main Lesson Book

The main lesson book, which should be kept with appropriate care in order to serve as a reference book for later blocks, can be subdivided into text, sketches, and maps. Texts are composed by the students themselves and in draft form should be checked and corrected by the teacher. The corrected version should be edited by the student according to the suggestions of the teacher and put into final form. Students independently decide on and prepare sketches, which complement their texts and were inspired through teacher sketches on the chalkboard. In contrast, maps should be prepared with great care during class time, following the instructions of the teacher, and referring to large wall maps that should be displayed in the classroom for an extended period of time. Special care should be used to completely label and annotate maps; maps without appropriate legends are useless. Thus the students develop an interest to create a clearly arranged and re-usable textbook on their own. By doing this, the student is able to connect him/herself with the material without the need to study it assiduously.

## The Block Contents

The technical basics of this Geology block are available in the book *The Tetrahedral Structure of the Earth*, Schmutz 1986. Besides numerous plates and illustrations, a bibliography is provided for further reading. Additional teaching aids are the large-scale demonstration sheets "Global Tectonics of the Earth—Earth's Tetraedric Shape." Please refer to the description of the grade 10 block "The Earth in Motion" for an examination of plate tectonics.

### *Earthquake Science [Seismology]*

During the decades following World War II, seismology played a key role in the study of the development of plate tectonics. The discussion of earthquakes serves as a good entry into the block, as it originates in what has already become and investigates the question of the process of coming into being. With the help of the portrayal of an historic earthquake, it may become clear that this process hidden in the interior of the earth involves the tearing-up and breaking-up of rocks.

The extraordinarily well-documented great earthquake that occurred in Alaska March 27, 1964, is a very vivid example. With its hypocenter at a depth of 20 km and its magnitude of 9.2, it caused ground displacements of more than 4 m on the earth's surface which corresponds to a displacement of approximately 15 m at the hypocenter. (National Research Council 1968, Schneider 1975, Bolt 2006) Due to the uplifting of mountains, it is possible today to investigate rocks on the earth's surface that were at the location of the earthquake processes in approximately 10 km depth when the alpine orogenesis occurred.

The crystalline rock of the Silvretta nappe [cap rock] in the Lower Engadine [Switzerland] and in Tyrol [Austria] show such traces of earthquake activity. (Schmutz 1995) Entire batches of rock look like they were torn apart and later on were perfectly cemented together by a dark mass. These types of rocks are called pseudotachylites (Fig. 2). Torn rock pieces show square and rounded forms; the dark mass consists of quenched natural glass, which developed from plasticized rock material melted from the enormous friction during earthquakes.

Now the hypothetical physical process can be discussed with the students. Under an overload of miles and miles, the approximately 250°C hot rock is subject to an even pressure from all sides. In addition, it is also subject to a pressure from the side, which attaches selectively to various portions of the rock. The rock bond resists until the directional pressure exceeds the resistance: All of a sudden the rock crushes. When rocks are torn into pieces at such depths, the developing cavities show the physical properties of an enormously strong vacuum. The abrupt sliding of rock strata in the scale of several meters' dislocation leads first to rock dust at the friction surfaces and subsequently to friction melting. Together with smaller and larger rock particles, the melted rock material is sucked up into the vacuum regions at lightning speed. The melt solidifies with such speed that allows the formation of natural glass: a pseudotachylite emerges (Fig. 2.2). After discussing this process until the students fully understand it, the question can be asked, whether it is conceivable that such an event can be caused only by pressure or also by tension—which would mean a tearing apart. The answer to this question is yes, earthquakes can be caused by directional pull [tension] as well as by directional pressure. Now when we take a closer look at the rock types of the Silvretta nappe, we also find rocks that are known as mylonites or ultramylonites close to the pseudotachylite zones. The structure within these rocks gives the impression of a flowing movement. Larger minerals [quartz or feldspar] are pulled out [dilated] into lenses and are located within an extreme fine-grained, but crystalline—not glassy—intermediate layer (Fig. 2.3). From rock mechanical experiments it is possible to understand the process of mylonite formation. If rocks contain a certain

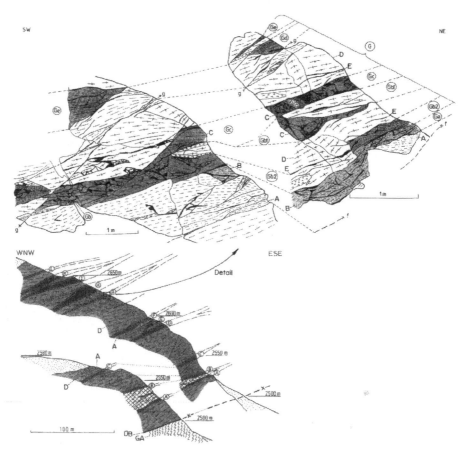

*Fig. 2.1:* Crystalline rocks from the Silvretta nappe [cap rock] in the Lower Engadine [Switzerland] ripped apart by the activity of ancient earthquakes and distorted along the plane of shear. A = Biotite gneiss, B = mylonitic biotite gneiss with pseudotachylite, C = zone of pseudotachylite breccia (comp. Fig. 2.2), D = amphibolite and hornblende-gneiss, E = pseudotachylite vein, f = strike shear plane, g = discordant shear plane, G = Eye gneiss, GA = Arosa green slate, ÜB = Silvretta overthrust, A to L circled = consequence of the pseudotachylite zones. (from Schmutz 1995)

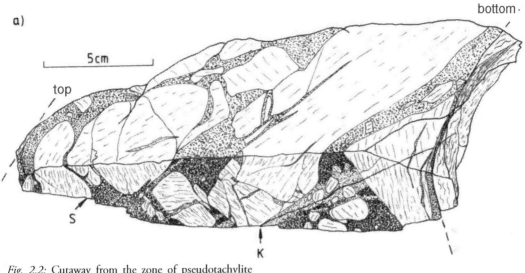

a)

bottom

5cm

top

S

K

*Fig. 2.2:* Cutaway from the zone of pseudotachylite breccia shown in Fig. 2.1. Square to rounded debris from amphibolites are embedded in the pseudotachylite matrix. a] half profile showing the cut end S at the front, b] bottom side, c] back, K = crack, S = cut. (from Schmutz 1995)

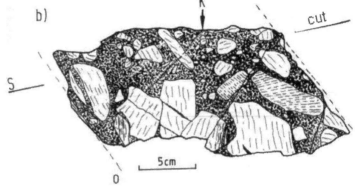

b)

K

cut

S

5cm

0

c)

0

u

5cm

*Fig. 2.3:* Cross section through a mylonite originating from granite. The larger grains are leftover feldspars. The striped bands consist of recrystallized, fine grinded material originating from feldspar, quartz and mica. (from Barker 1998)

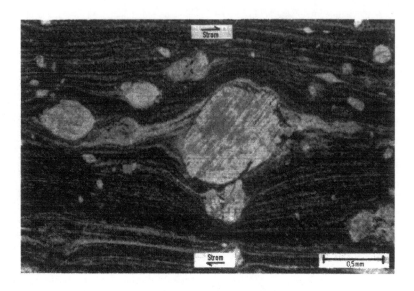

amount of water, directional pressure can continuously influence the rock and a slow sliding shear movement will occur layer after layer, accompanied by a selective reduction of the grain size.

No earthquakes result if in the depth the directional pressure can be transformed into continuous movements. The process of an earthquake can therefore be understood as a movement that has been suppressed for a long time and is now caught up on in a very short time. In addition, students take note of the fact that movements in the earth's interior are caused not only by earthquakes, but also continuous movements, as in the case of the mylonite formation.

After the detailed examination of these rock types, which introduces the students to the process of rock metamorphosis, we can now study the worldwide distribution of earthquakes using a distribution map of all recorded earthquakes. But before we get too far into this, students usually want to know how earthquakes are recorded. There are some good books describing the construction and function of earthquake recording devices, e.g., seismometers. (Brinkmann 1990; Strohbach 1990; Shearer 2009) Students are usually very satisfied when they fully understand the function of the measuring devices. The analysis of seismograms from various areas of the world should be explained in a simplified form. It is well worth discussing the phenomena of refraction and reflection at the interface boundary between different media; however, characteristic features of longitudinal, transversal and surface waves can be demonstrated only in a reduced form. One significant detail is how seismologists assume hypothetic interface boundaries inside the earth's interior in a trial-and-error way until the sum of all seismograms points to a particular hypocenter. What comes out of this? As a side effect this leads to an image of the architecture of the earth's interior!

The map of the world showing all earthquakes registered in a decade deserves to be investigated in more detail and should be carefully copied into the main lesson book (Fig. 2.4). At first glance, the absence of earthquakes in most of the interior of the continents and a line of earthquakes in the center of the oceans make for striking observations. In the Atlantic Ocean, starting in the Arctic Sea, earthquakes are concentrated in the center and branch out in

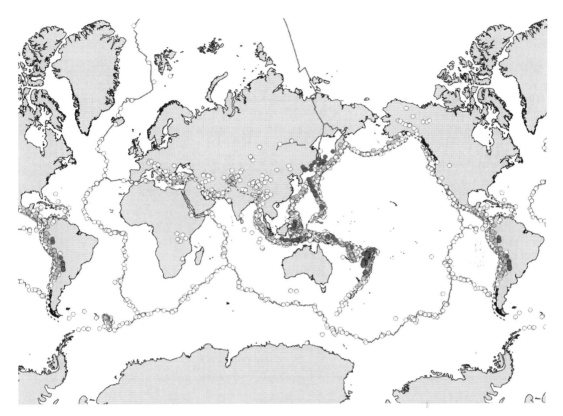

*Fig. 2.4:* Map of the worldwide distribution of earthquakes between 2000 and 2006 with a magnitude above 5.5. Clear dots = shallow earthquakes with a focal depth of 1–100 km, gray dots = intermediate quakes with a 100–300 km depth, solid black dots deep quakes have a 300–700 km depth. Shallow earthquakes occur along plate boundaries (solid lines); Earthquakes with increasing depth (clear-gray-black dots), the so-called Wadati-Benioff zones, occur at convergent plate boundaries (dashed line). Map prepared by the translator using publically available data from the National Earthquake Information Center (NEIC), plotted with the use of a GIS system.

the south Atlantic Ocean. The margins of the oceans are mostly free of earthquakes. The seismic foci appear in a line, pointing to the center of the earth.

In the Pacific Ocean the very opposite is the case. Most of the earthquakes are located on the margins of the ocean. The focal planes are pointing from the oceanic trenches slanting down under the continents and island arcs. The arched distribution of earthquakes in the south Pacific Ocean nestles in California against the slantingly submerging distribution of earthquakes. The Indian Ocean shows a balancing intermediate situation; it connects the polar appearances of the Atlantic and Pacific Oceans. In terms

of earthquakes, the west Indian Ocean still shows Atlantic character whereas the NE Indian Ocean shows Pacific character. Like the mediation character of the Indian Ocean, the motif of the polarity of the Atlantic and Pacific Oceans will reappear several times again. Whereas the earthquake belts in the Pacific Ocean are located at the continental margins, slantingly submerging to a depth of 700 km, the situation is very different in Europe and south Asia. The earthquake belt separates Africa from Europe. If India is seen as a geologic subcontinent, the Tibetan earthquakes are also located on the margin of a continent. According to newer research (Pavoni 1997; Pavoni and Müller 2000), the radially-oriented mid-ocean earthquake zones are mainly caused by the earth's crust being pulled apart [tension phenomena]. The slanted submerging earthquake areas of the continental margins are caused by compression phenomena. This polarity of compression/tension can be schematically summarized as follows:

| Compression | Tension |
| --- | --- |
| Continental margin | Oceanic center |
| Pressing together | Tearing open |
| Slanted submerging | Radial arranged |
| Focal depth 0–700km | Focal depth 0–700km |
| Broader field | Narrow zone |
| Sporadically even, strong earthquakes | Mostly weak to moderately strong earthquakes |

## Volcanism

As the activities of volcanoes are closely related to the processes of earthquakes, there is a perfectly smooth transition between these topics. Again, it is possible to start with volcanic rocks, which constitute the solidified volcanic processes, observe the superficial phenomena and elaborate the processes to volcano formation. To follow the polarity of pressure/tension, a subdivision into andesitic and alkaline basaltic volcanoes is the thing to do.

### Andesitic Volcanism

The exemplary portrayal of a volcanic eruption is again a favorable introduction. The eruption of Mount St. Helen's is a prime example, as there is plenty of good visual material available. Other choices could include the monumental explosions of

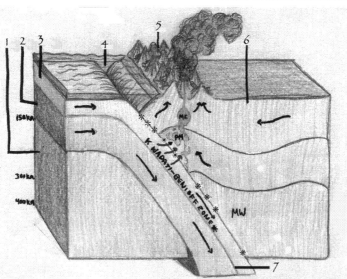

*Fig. 2.5a:* Whenever dense but thin oceanic crust (3) collides with less dense but massive continental crust (6) in a convergent plate boundary, the heavier and thinner oceanic crust is bent downwards and starts to dive below the thicker and lighter continental crust. This downwards movement of the oceanic crust creates the deepest depressions of the earth's crust called oceanic trenches (4). The colliding continental crust is bent upwards and mountains are formed (orogenesis). Increasingly deeper earthquakes occur along the sliding surface between the two colliding plates (✳ = shallow earthquakes, ✴ = intermediate earthquakes, ✻ = deep earthquakes). This process is called subduction. In honor of its discoverers, this earthquake pattern is named a Wadati-Benioff zone. It is shown here in the case of the west coast of South America. In a depth of 100–300 km, the increased temperature and pressure cause water and carbon dioxide to be forced out of the subducing oceanic crust (blue arrows) of the slab (7). This causes the rocks of the lithosphere (2) and the mantle wedge (MW) to melt at a relatively low temperature and create magma (partial melting, PM), which rises and gathers in magma chambers (MC) feeding stratovolcanoes in between and on top of the Andes (5). Drawing by Madelaine Bradford

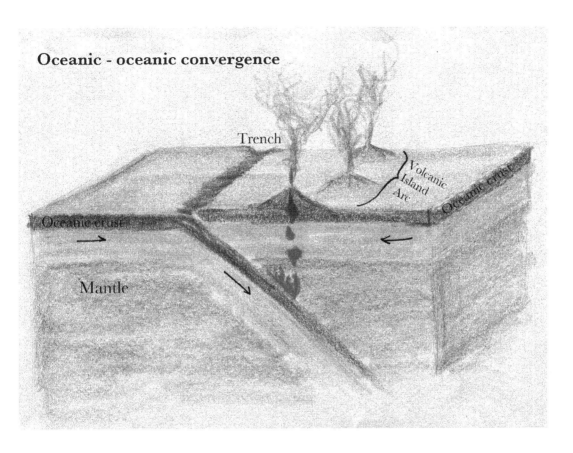

**Oceanic - oceanic convergence**

Trench

Volcanic Island Arc

Oceanic crust

Oceanic crust

Mantle

*Fig. 2.5b* Whenever two dense and thin oceanic crusts collide in a convergent plate boundary, the heavier and thinner of the oceanic crusts is bent downwards and starts to dive below the thicker and lighter oceanic crust. The downwards movement of one of the oceanic plates underneath the other also creates oceanic trenches and Wadati-Benioff zones and causes partial melting. The resulting magma chambers feed stratovolcanoes that rise above sea level and create a volcanic island arc. Drawing by Maria Hagen

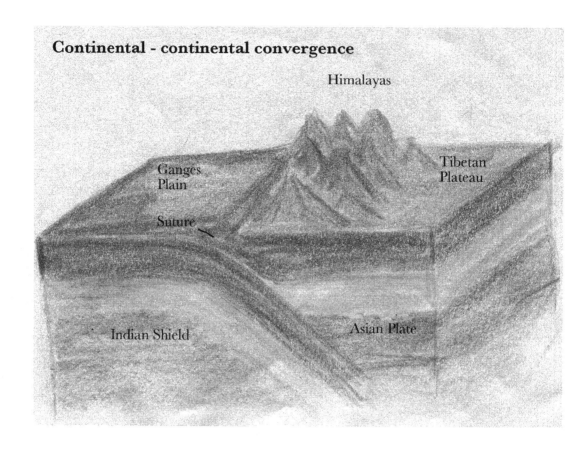

# Continental - continental convergence

Himalayas

Ganges
Plain

Tibetan
Plateau

Suture

Indian Shield

Asian Plate

*Fig. 2.5c* Whenever two less dense and thick continental crusts collide in a convergent plate boundary, the heavier and thinner of the continental crusts is bent downwards and starts to dive below the thicker and lighter oceanic crust. The downwards movement of one of the continental plates underneath the other creates a downward facing seam (suture) but no Wadati-Benioff zones and no partial melting—thus no stratovolcanoes. The collision results, however, in the formation of mountains on the lighter and more massive continental plate, as shown here of the collision between the heavier Indo-Australian plate forming the Ganges Plain and the lighter and thicker Eurasian plate that is folded into the Himalayas and the Tibetan Plateau. Other continental collisions can be almost even and cause orogenesis on both colliding plates (example: collision between the African and European plates forming the Alps). Drawing by Maria Hagen

Krakatoa or Mount Tambora in Indonesia, or the historically interesting eruption of Vesuvius and the destruction of the Roman cities of Pompeii and Herculaneum (compare to Francis and Oppenheimer 2004; Johnson 2006; Pichler 1988; Rast 1987; Schminke 2003; Volcano, the eruption of Mount St. Helens, 1980). The characteristics of andesitic eruptions are a long dormancy and a very vaguely announced but then violent activity, which leads to a rhythmic alternating deposit of ejected volcanic tuff and run-out lava along relatively steep mountain slopes [stratovolcanoes or composite volcanoes] (comp. Fig. 2.7b). For more specific information, please consult the literature cited above.

Consideration of the architecture and origin of the magma chamber will lead to a better understanding of these impressive events. Andesitic magma chambers are located at a structure which we already discussed, namely close to the earthquake zones which are slanting, reach down below a continent and are subjected to the interplay of forces that push the earth's crust together. In a depth of 60 to 70 km, this zone is almost free of any earthquake, as this is the depth where magma chambers occur, which consist of liquid, and therefore unbreakable, rock material. Estimates of the temperature in the earth's interior show that the areas where the partially melted rock occurs are characterized by the lowest relative

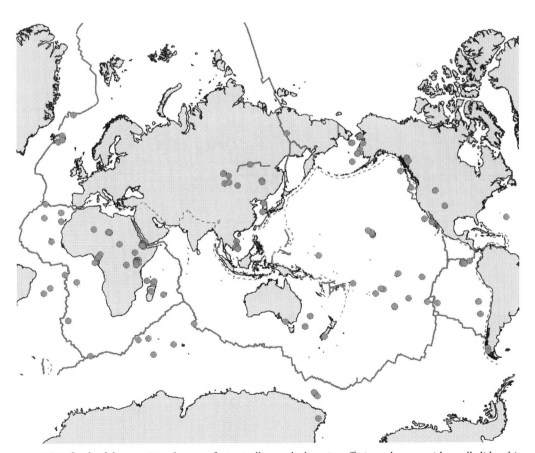

*Fig. 2.6: Left side of the map:* Distribution of seismically very little active effusive volcanoes with an alkali basaltic to tholeitic chemism (gray dots). Volcanoes known as "hot spots" break either through the continental crust or are located in rift zones or they break through the oceanic crust. Solid lines show divergent plate boundaries, dashed lines show convergent plate boundaries.

temperatures! This contradiction needs a more detailed explanation. It is known from laboratory experiments on the liquefaction of rocks with a granite chemism that a granite with a certain water content within its pores and hairline cracks can have a melting point as low as approximately 700°C. Well-drained and desiccated but otherwise identical granite, in comparison to this, does not melt below 1300°C. If, therefore, rocks in the region of the andesitic magma chambers would show increased water content, this would clear up the contradiction (compare to Schminke 2003).

In anticipation of what will come later in the course of the block, the phenomenon of the subduction zones should be mentioned here. The rocky oceanic crust of approximately 10 km thickness submerges under the continental crust, which is approximately five times as thick. The interface boundary of this motion coincides with the above-mentioned earthquake zone at the continental margins (see Fig. 2.5a). The zone of submersion, originally discovered by the Japanese seismologist Kiyoo Wadati and later further described by the American seismologist Hugo Benioff, starts at an oceanic trench and can be followed (indicated by the earthquake zone] to a depth of more than 700 km slanting below the continent. Oceanographers have known for a long time about the intensive

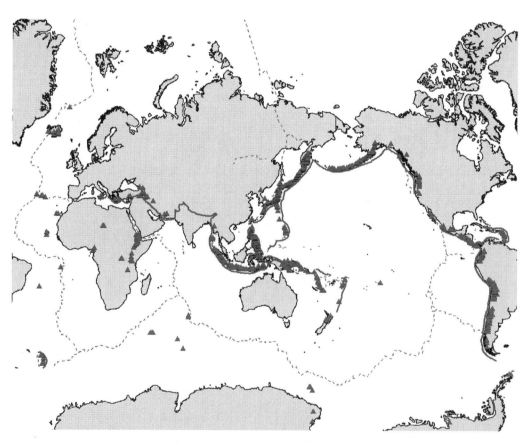

*Fig. 2.6: Right side of the map:* Distribution of the acidic, andesitic to dacitic volcanoes (triangles). Seismically highly active volcanoes with a carbonate-alkaline to andesitic chemism are predominantly located in relatively young, folded rock portions or, more precisely, at subduction zones. Solid lines show divergent plate boundaries, dashed lines show convergent plate boundaries. Map prepared by the translator using publically available data from the Smithsonian Institution, Global Volcanism Program, plotted with the use of a GIS system.

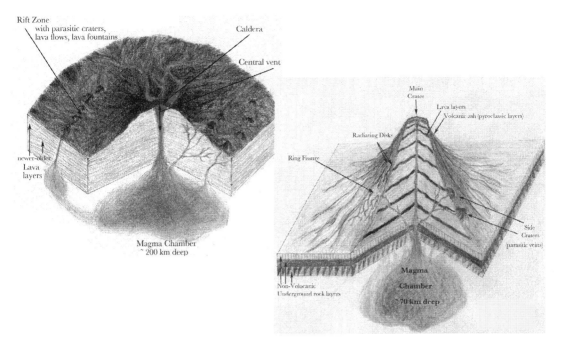

*Fig. 2.7:* Comparison of shield volcanoes (left) and stratovolcanoes (right). Shield volcanoes occur mainly in oceanic areas and are only slightly inclining, flat but nevertheless often quite high volcanoes, built up from layers of lava flows with an alkaline-basaltic chemism. Strato- or composite volcanoes occur exclusively in subduction zones; their much steeper flanks consist of alternating layers of pyroclastic material and lava flows. Drawing by Ruby An based on Rast 1987

life, especially of the inconspicuous single-celled phyto- and zooplankton, in the coastal waters above oceanic trenches. This richness is mainly caused by the upwelling of nutrient-rich and cold hypolimnic water combined with strong insolation. When there is plenty of life, there is also plenty of die-off. Vast amounts of skeletal remains of planktonic organisms are sinking towards the deep sea, mostly degrading and dissolving, and chemically precipitated in the form of lime after supersaturation occurs. This lime sludge is accumulating within the oceanic trench and is slowly transported into the depth by the submerging of the ocean floor rocks. In this way, it is reasonable to believe that in a depth of 60 to 70 km (corresponding to the location of andesitic magma chambers),

excessive amounts of water and carbon dioxide occur in the vapor phase, which will lead to the above-mentioned lowering of the melting point of the rocks.

Sun—water—life—death—lime slurry—subduction and metamorphous transformation of minerals—partial melting—ascend—volcanic eruption: This chain of cause and effect will get an even deeper meaning in the cause of the earth science curriculum (comp. chapters 3 and 9].

After the students understand the existence of the three-dimensional tube-like, sometimes expanding and sometimes thinning-out magma chamber, it is time to discuss the ascending of the masses. As there is sufficient water available and caused by a high vapor pressure, the igneous melt, nourished from rocky material of the continental crust and a smaller amount of ocean floor rock may get quite movable. The magma ascends along cracks and zones of material weaknesses and adds rocky material that lies on its way into the melt. As the temperature gradient and differences in hardness relative to the neighboring rock rise as further up the melt moves, it often gets stuck in large amounts in the depth of several kilometers and starts to cool off very slowly. This is the way plutonites or intrusive rocks such as granite, granodiorite, syenite and diorite are formed. Only if the last breakthrough to the surface succeeds, accompanied by earthquakes and fold formation, do we arrive at the above-described phenomenon of a severe volcanic eruption.

Now when we draw a world map of all active or "sleeping" andesitic volcanoes (see Fig. 2.6), a familiar pattern reappears: the same pattern as for the compression earthquake zones. The spatial geometry of these zones will be discussed at a later time.

### Alkaline-Basaltic Volcanism: Hot Spots

The volcanic phenomena on Iceland, Hawaii, at the Etna [Sicily, Italy] or long time extinct volcanoes like Yellowstone or the Eifel in Germany are showing other, completely different characteristics. At the Etna, for example, eruptions occurring along ripped-open fissures are reported every other year. The main activity is the outflow of lava. The corresponding mountain form is much less steep-conic as compared to andesitic

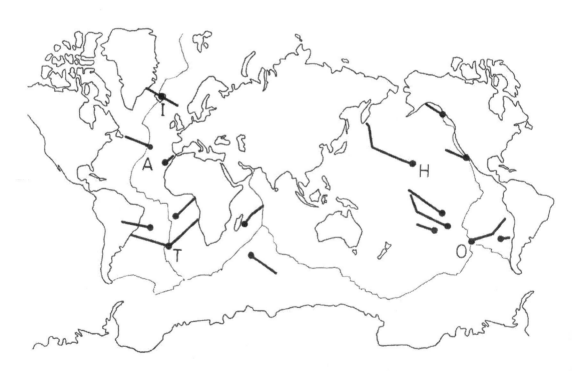

*Fig. 2.8:* Important hot spots and the associated strings of islands or submarine basalt ridges. A = Azores, H = Hawaii, I = Iceland, O = Easter Islands, T = Tristan de Cunha. (according to Schminke 1986, Schminke 2003)

volcanoes (Fig. 2.7). Volcanoes created by alkali-basaltic lava in such a way are called shield volcanoes. Their rock types belong to the chemism of the earth's upper mantle; they are therefore basic/alkaline, which means that free silica does not occur as quartz and there are only small amounts of alkaline varieties of feldspar. Main minerals are the dark colored olivine, pyroxene and hornblendes. A closer look at the islands of Hawaii leads us to an important chapter on how the idea of plate tectonics was developed. In the 1960s, a lining up of volcanoes into a 6000 km-long row was discovered, reaching from Hawaii to the Bering Sea/Kamchatka, with most volcanoes now eroded below the surface of the sea. It could be shown with a clear regularity that the extinct volcanoes are older and older the further away they are located from Hawaii.

This row of volcanoes and their age structure hint at a cyclic volcanic process that must have been active for more than 90 million years. The melting zones are expected to be in a depth of approximately 700 km. These zones located deep within the earth's mantle in alkaline-basaltic material are commonly referred to as hot spots. The Canadian geophysicist J. Tuzo Wilson was systematically looking worldwide for hot spots based on the known structure aligned with rows of volcanic cones (compare to Giese 1987, Wilson 1976). Within the Atlantic and Indian Oceans, he discovered a series of symmetrical double rows of strings of volcanoes, which on both sides of the middle of the ocean stretch from an active volcanic island through the entire half of the ocean until they reach the flanking continents (Fig. 2.8). This is most obvious with Tristan de Cunha in the southern Atlantic and Iceland in the northern Atlantic. Students will quickly make up their minds about this:

That two hot spots come together in the middle of the ocean cannot be random. The same phenomena and their thoughtful appreciation led Wilson in 1962 to the drafting of the theory of plate tectonics. He assumed that hot spots are stationary structures within the earth's mantle, and the earth's crust [astenosphere], split up into plates, would move above these hot spots. The plates would move sideways away from the already discovered mid-ocean ridges; the improbable coincidence is transformed into an inner necessity. The appreciation of this scientific result leads students to an experience of trust into their own thinking, as a law of nature [causality] is grasped in the act of thinking by relating it to an experienced necessity.

From the knowledge of the development of regions of partial melting in andesitic volcanoes develops the question: Why do hot spot melts occur? From seismological research it follows that mobile ascending material from the transition zone between the earth's lower and upper mantles [the so-called mantle plume] accumulates in isolated or serial areas of melts [hotspots]. When the very slow melting into the rock above continues, the hotspot material forms a new magma chamber, where the mobile liquid stays for a prolonged time, when it comes to a pressure release from only having approximately 5 to 10 km depth of rocky cover above the melt. (Schminke 1986, 2003) A rhythmic opening occurs, as described for the hotspot volcanoes, when the magma is able to flow out through stress zones within the earth's crust. We can therefore describe the entire volcanic processes in terms of just four depth zones: the zone of [partial] melting, the more extended zone of detachment and ascending, a stagnation and differentiation zone close to the surface, and the eruption zone visible to direct observation.

## The Mid-Ocean Ridges [MOR]

A look at a relief map of the ocean floors (Hess 1962) stimulates questions about the formation of such concise mountain ranges, which spread predominantly along the center of the oceans. When watched as a cross-section, this largest of all structures on earth shows a remarkable similarity to alkali-basaltic volcanoes. Both rather flat mountain flanks do not lead to a peak but to an area of subsidence which is similar to the calderas of volcanoes. The entire ridge is formed by alkali-basaltic magma, thus showing again a close relationship with hotspot volcanoes. The course of the ridge, named MOR, is known to the students from the interpretation of the earthquake map (Fig. 2.4); it is identical to the regions of oceanic earthquakes, which reach down to about 70 km depth and point radial to the center of the earth. They are caused by the tearing up or the stretching [dilation] in the earth's interior.

The investigation into these mid-ocean ridges is again an exciting chapter in the history of the earth sciences. During the survey of the ridges, scientists also recorded the field intensity of the earth's magnetic field. These results are another key to the mystery of the MOR. What was found is an alternating sequence of abnormally intensified or abnormally weakened field strength in more or less similar, parallel and laterally symmetric regions (Fig. 2.9). Based on analyses of rock samples from both sides of the ridge, it was discovered that the elongated crystals of the mineral magnetite, which occur almost always within the alkaline basalt, were lined up with the earth's magnetic field at the time when they first crystallized from the melt.

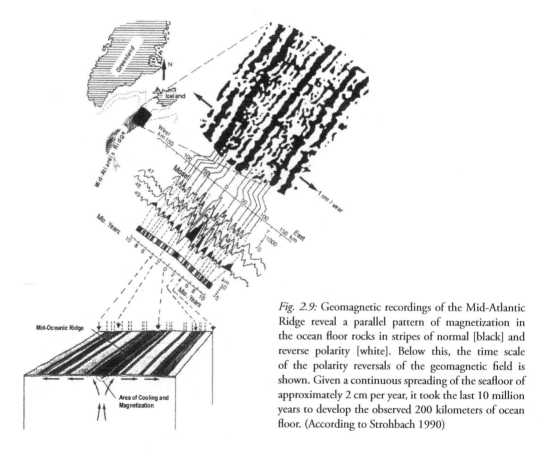

*Fig. 2.9:* Geomagnetic recordings of the Mid-Atlantic Ridge reveal a parallel pattern of magnetization in the ocean floor rocks in stripes of normal [black] and reverse polarity [white]. Below this, the time scale of the polarity reversals of the geomagnetic field is shown. Given a continuous spreading of the seafloor of approximately 2 cm per year, it took the last 10 million years to develop the observed 200 kilometers of ocean floor. (According to Strohbach 1990)

This anomaly can be explained by a periodic reversal of the North and South Poles of the earth's magnetic field. Thanks to the fact that the rocks of the deep sea all consist of the above-mentioned alkaline basalts and that the age of these pole reversals can be dated using additional rocks from hotspot volcanoes which also originate from the ocean floor, it became possible to determine the crystallization age of more than half of the earth's crust in a very short time (Fig. 2.10). To do the same task for the continental crust is very time-consuming and we are, even today, far from completing this job.

With the help of these facts, the breakthrough to a general acceptance of plate tectonics was almost achieved. The only thing left to do was to clarify how rocks are formed in the long central rift of the MOR. It is possible to describe this event to the students by falling back on the activity of the hotspot volcanoes and reminding them that even huge amounts of ascending masses can get stuck just before reaching the surface and form intrusive rocks because they cool off slowly. Most of the slowly ascending, plastic-viscous magma from a melting region in 700 km depth gets stuck under the central rift valley and forms crystalline peridotite. Rarer are outflows close to or at the surface, which occur when the central rift valley collapses abruptly along fractured surfaces. This gives rise to gabbroid basalts. Even rarer are mostly unrecognized eruptions, which resemble the effusive eruptions typical of alkaline basaltic volcanoes. In this case, basalts with a structure of interstratifications and the typical pillow lava are formed as shock-like cooling phenomena from cold ocean water.

This triad of peridotites, gabbros and pillow lava built the entire region of all deepsea floors up to a depth of approximately 10 km. This triad of rocks usually cannot be

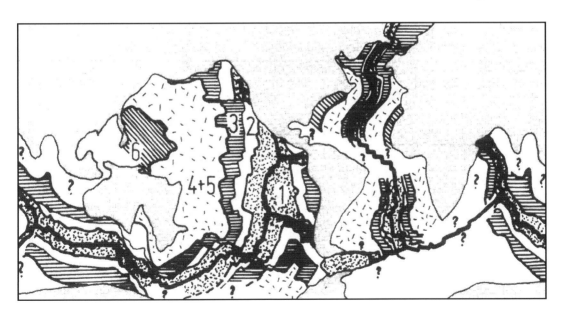

*Fig. 2.10:* Age structure of the seafloors determined according to the distribution of magnetic anomalies. The continental crust [continents and shelf seas] is shown in gray. 1 = 0–20 Mio. years, 2 = 20–40 Mio. years, 3 = 40–65 Mio. years, 1–3 = Tertiary, 4–5 = 65–135 Mio. years [Cretaceous], 6 = 135–190 Mio. years [Jurassic]. (From Schmutz 1986, data by Muller, R.D., M. Sdrolias, C. Gaina, and W.R. Roest 2008)

observed as they occur under large bodies of water and mud. They can, however, be observed in folded mountains, which originate from a subduction zone where portions of seafloor rock were flaked into continental rock and were subsequently lifted to the surface of the mountains by uplift. Exemplary investigations into this phenomenon took place in Oman, on Cyprus and in the Californian coastal range. Easily accessible examples are the mafic and ultramafic rock complexes close to Chiavenna in northern Italy. (Schmutz 1976)

## Continental Rift Valleys

In two locations of the earth, alkaline-basaltic hotspot volcanoes are coupled with large-scale rift valley structures: in central Africa and the Baikal region in central Asia.

In addition to events of collapsing walls, the flanks of the rift valleys show a drifting apart [spreading] of its flanks at about 1 to 3 cm per year. Seismic investigations show that the lower asthenosphere is thinned out at these locations, resulting in the fact that mantle rocks reach up very high below a shallow crust. It looks like there will soon be a separation recurring, resulting in a new ocean with its typical seafloor rocks. This is actually happening at the Red Sea, where the mid-oceanic ridge in the western Indian Ocean reaches into the old African continent. These areas, which will receive another significance during the earth science block in grade 12 (chapter 9), show the character of spreading [dilatation] just like a typical MOR. (Schmutz 1996; Suchantke 1993, Suchantke 2001)

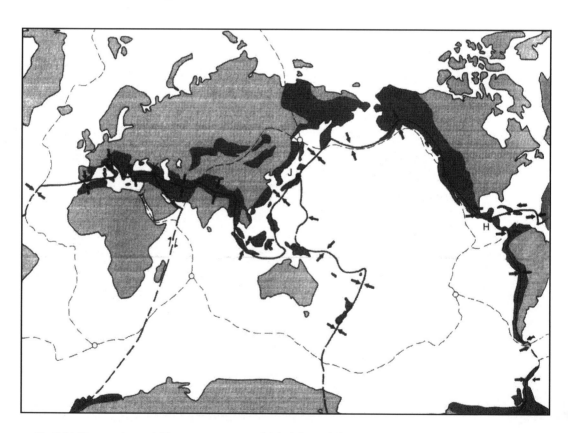

*Fig. 2.11:* Compression and dilatation zones on earth The left panel shows compression zones, the right panel points out zones of expansion [dilatation]. Arrows indicate relative plate movements on plate boundaries. K = Caucasus,

## The Compression and Dilation Zones of the Earth

As a preliminary summary of the block, the discussed results can be brought into a pattern which illustrates the worldwide phenomena. In order to do this, we return to and expand on the pattern from earlier in this chapter.

When we now draw the compression zones of the earth onto a globe (Fig. 2.11, left), the spherical surface is divided into four areas. Three areas meet at four sites on earth: Honduras, the Caucasus mountain range, Japan and the South Pole. (Schmutz 1986; Steiner 1924a) The line of compression between the Caucasus and the South Pole shows an untypical form due to the fact that it mainly runs through the complexly structured and not sufficiently understood Indian Ocean.

Surprisingly, we also receive four components of the earth's spherical surface when we visualize the lines of dilatations on the globe (Fig. 2.11, right). The four sites of conjunction of three partitions are located in the south Atlantic, central Indian Ocean, in the southeast Pacific and at the peninsula of Kamchatka.

It is an important realization when students discover that this second discovered four-field structure is an upside-down image of the previously discovered counterpart. However, the double tetrahedral structure of the earth (Schmutz 1986) should be discussed further only if students themselves discover this symmetry on the globe. It is otherwise

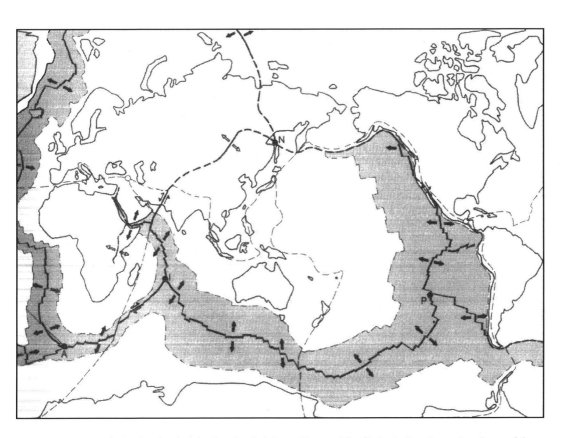

J = Japan, H = Honduras, S = South Pole, A = South Atlantic [Bouvet Island], I = Indian Ocean [southeast of the Mascarene Plateau], P = South Pacific [Easter Islands], N = Northeast Siberia. (From Schmutz 1986)

more beneficial, to introduce the polarity of the tetrahedron in the crystallography block in grade 10.

The experience that the lines of compression and dilatation obviously contribute to a meaningful geometric unity is a very important aspect in grade 9. Students foresee or recognize with an inner satisfaction that the earth is geometrically structured so that the ratio between tension and pressure is in balance and becomes visible as the symmetry in the earth's shape.

The conceptual grasping of the balance between tension and pressure as a physical law is achieved in the physics block on static in grade 10. In the geology block it is more important to allow for the experience of the meshing and interlocking of both polar phenomena. It is of particular interest to investigate certain sites in more detail.

In California, for example, the normally separated lines closely meet for a length of several thousand kilometers. The result of such a synergy of tension and pressure is the San Andreas Fault. Along this earthquake zone, the Pacific part of California is shifting to the northwest relative to the interior of the continent at an average rate of several centimeters per year. Whenever the motional opposites of expansion and compression are meeting, a new, third mode of motion appears, the so-called horizontal displacement or strike-slip fault (see Table 2.1).

### General Tectonic Processes and Petrology

If you followed the didactic sequence of this block so far, the questions may have arisen as to where and when students will learn about principle tectonics and geologic

| Compression zones of the Earth | Dilation zones of the earth |
|---|---|
| *Earth quakes* **on the continental margins** | *Earth quakes* **in the center of the oceans** |
| **Slanted Submerging until 700 km depth** | **Pointing radial around the center of the earth until 70 km depth** |
| **Crushing due to being pushed together (Pressure)** | **Crushing due to being pulled apart (Tension)** |
| *Andesitic volcanoes* **with magma chambers in subduction zones in depths of 70 km** | *Alkaline-basaltic volcanoes* **with magma chambers in depths of 700 km (hotspots)** |
| **Granodioritic to granitic chemism** | **Alkaline-basaltic chemism** |
| **Explosive eruption accompanied by strong earth quakes** | **Lava outflow with only moderate eruptions** |
| **Formation of steep stratovolcanoes** | **Formation of much flatter shield volcanoes** |
| **Subduction zone starts at typical** *oceanic trenches* **close to coasts and moves below the** *high mountains* **of the earth, which are still uplifting.** | *Mid-oceanic ridge (MOR)* **with a continuously collapsing central rift zone stretches along the middle of the oceans or forms** *continental rift valleys* **when reaching inside of continents.** |

**When compression and dilation come together, the result is a horizontal displacement (transformation) of portions of the earth's crust**

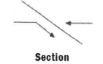

**Section**          **Flat (planar)**

**Flat (planar)**

*Table 2.1*

processes, and where and when general petrology and mineralogy are covered. This means understanding the formation and back formation [dissolution, erosion] of the rocks (rock cycle) and also the processes of folding, breaking, sinking and uplifting of rock masses. The concept of the geosyncline should be developed.

This block is designed to incorporate the mentioned topics when and if they are needed to discuss volcanism, earthquakes, compression, dilatation and mountain formation [orogenesis]. As there is excellent introductory literature available (Brinkmann 1990, Grotzinger 2010), these concepts are not described here any further. There is also a book available which illustrates these topics with exemplary geological excursions [field trips] into the Alps. (Schmutz 2005)

Petrology/mineralogy should also be embedded into the storyline of this block instead of being taught as an isolated discipline. It is essential to have a good rock collection consisting of large, illustrative example rocks.

## A Forward Look to Plate Tectonics

If the observation and understanding of phenomena was developed carefully and there was substantial involvement of the students themselves in this development, then four weeks are [almost] over by now. If time permits, the consequences of the newly learned material can be studied in a few selected examples.

Best suited is the African plate. It contains Africa with the Mediterranean, the southeastern Atlantic and the southwestern Indian Ocean including Madagascar. Except for the mountain chain of the Atlas, Africa does not show any compression phenomena. If the new earth crust is continuously produced along the mid-oceanic ridges around Africa, this must lead to an increase of the plate. The shape and course of the MOR, which resembles a blown-up triangular form similar to the shape of the African continent, shows that this must have been the case in the past. However, this imagination results in the necessity for South America, Antarctica and Australia to move away from Africa, which, furthermore, means that the regions on the other side of the globe are subject to substantial phenomena of compression, if the earth should maintain approximately the same size. If the triangle becomes larger with its side towards the South Pole, it will be slightly pushed towards north due to the restriction from other parts of the earth. In this way, there must be compression phenomena occurring at the boundary with Europe, which is, as a matter of fact, the case. This way, we could develop a first approach leading to an understanding of the Alpine orogenesis (Fig. 2.12a and b).

This example shows that the imagination of plates being pushed back and forth is inappropriate. Plates change constantly in shape and size. They grow at dilatation sites, mainly at the MORs. Caused by these continuous changes in shape, the distance between the passively pushed continents diminishes or enlarges as a logical consequence. However, this discussion takes 9th graders to their limits in imagination and comprehension. Therefore, it is advised not to continue with this topic before grade 10, when the grasping of constant changes in shape and the consequences of such phenomena will be the main content of the block. This is the reason that the dynamic processes of plate tectonics should be given enough time in grade 10.

*Fig. 2.12a:* The lack of compression zones at the western and eastern shores of Africa leads to a compression in the Mediterranean and the Alps against Europe. This is due to an attachment of oceanic crust along the MOR around Africa, which pushes the African northward wedge.

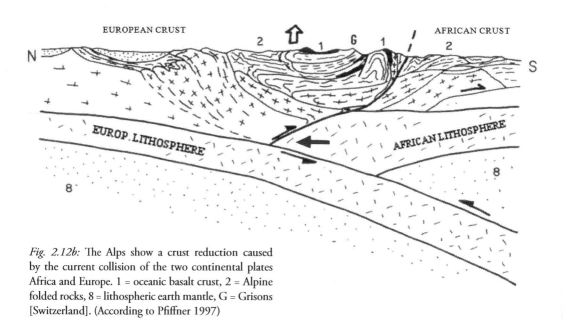

*Fig. 2.12b:* The Alps show a crust reduction caused by the current collision of the two continental plates Africa and Europe. 1 = oceanic basalt crust, 2 = Alpine folded rocks, 8 = lithospheric earth mantle, G = Grisons [Switzerland]. (According to Pfiffner 1997)

# Chapter 3

# The Earth in Motion—Tenth Grade

## The Nature of the Task to Teach Grade 10 According to the Study of Man

There is no other grade that offers more earth science in high school than grade 10. The history blocks originate from earth science topics. The old cultures should be introduced to the students in such a way that the geographic-physical-topographic conditions of the regions relate to the corresponding cultural impulses. In chemistry, earthy solids such as metals and salts are discussed; it is like teaching some kind of chemical mineralogy. In the surveying block, mathematics is taught in relation to the earth. In its own block, earth science in grade 10 covers classic geography as "The Earth in Its Entirety" or "The Earth in Motion." The meaning and mode of realization of the also possible crystallography block is presented in chapter 4. Chapter 6 offers some suggestions as to what extent the life skills and technology lessons can become increasingly more related to the earth sciences when progressing from grade 9 to 10. As I mentioned before, Steiner stressed, in a lecture about the education of adolescents explicitly, that there has to be an abrupt transition from grade 9 to grade 10 in order that the students do not simply continue to gain knowledge but begin their struggle for understanding. (Steiner 1922a; comp. chapter 2)

In the transition between the ages of 14–16, the soul of a young person requires exercise to practice and shape the power of making judgments, which faculty is now experienced more consciously. Students want to make judgments themselves by activating and training their ability to think dynamically. Pure causal judgments in the realm of the inorganic are completely determined by the matter/the thing itself. However, when it is the task to recognize motion as itself, the inner formative forces of the human soul are in much higher demand. For the student it is not longer important to ask about what and how of a matter/thing, but for the why. When the topic is motion, it is especially exciting to ask the why. In this way the students learn to distinguish between spontaneous or active motion and being moved [passive or forced motion]. This leads to an understanding of the difference between the world of the lifeless inorganic and the world of the alive organic also within the method of judgment. (Steiner 1886, chapters 15 and 16) In this way it leads inevitably to the latent [dormant, concealed] question: Does Earth in its entirety constitute an organism or is it a dead inorganic structure?

## Themes of the Block "The Earth in Motion"

The the latent question mentioned above is the theme and runs all the way through grades 10 to 12. In the earth science block of grade 10, students should begin to work on this question properly and subtly. This is achieved by focusing on a variety of motions in the air, water and rock spheres of the earth [atmosphere, hydrosphere and lithosphere]. It is important to create the relationships between pulsating vertical motions and circulating horizontal motions. I am inclined to call this block "Earth in Motion" rather than "The Earth in Its Entirety," as in grade 10 the connection of the cosmos and the human being as an acting cultural being is indicated only to a limited extent. It is not before grade 12 that the earth can be appreciated as a single whole entity.

As there are vast quantities of complex subjects within physical geography (Rohrbach 1999), it is suggested to focus on a few exemplary dynamic phenomena to allow for a controlled practicing of the formation of judgments. This provides time to allow students to deepen their experience of self-created judgments. As a preliminary stage to this block, it is advisable to present the crystallography block first. This will lead to the experience of a living concept, or in other words, it will lead to the interior activity of a metamorphosis of thoughts by means of the seven crystal systems. In order to achieve this, the ability to imagine spaces and the appreciation of symmetries are enhanced through this preparatory block. Both are important preconditions for the understanding of the motion phenomena of the earth. Compared to the geology block in grade 9, in grade 10 it is all about overcoming too narrow interpretations of the causality principle. In order to penetrate motion phenomena, purely physical explanations are no longer sufficient.

The firm ground of the causal-logical formation of judgment in the inorganic world should not be retracted abruptly from the students. On the contrary, it should be worked towards an extension of the ability to judge, where the non-causality of the inorganic is cautiously complemented by the causality of life processes. The target, to safely fight your way through a recognizable understanding of the moving and living world, cannot be approached more clearly and carefully. The almighty natural sciences nowadays are actually facing the same threshold. When jumping too fast from the lifeless to the spiritual, one runs the risk of not appreciating the spiritual enough. It is exactly the sequence of steps from the Lifeless to the Living and the Soulful to the Spiritual which needs to be grasped in the process of thinking (chapter 1).

## Canon of Subjects

Two modes of motion shall be investigated in three partitions of the earth. It is advisable to present these connected motions in a differentiated way, as this connectedness changes substances and qualities. This way one does not fall back into the rigid imagination of a mechanical circle, in which there occurs at best only feedback mechanisms. The first one is about the flow, where a horizontal component dominates over a rising and descending motion. The second mode of motion can be characterized by staunching and loosening; this mode is dominated by a vertical movement. Both modes of motion are certainly interdependent but should at first be investigated separately. In the seventh lecture of the third scientific course for teachers, Steiner points out how important it is to distinguish between the horizontal position of the animal spine, together with the horizontal movement of the animals, and the vertical gesture of plants. The human being corresponds to a resultant between the horizontal and the vertical. (Steiner 1921) In this current block we work with this qualitative differentiation between "horizontal" and "vertical" in order to achieve the preconditions for the students to distinguish qualitatively between plant, animal and human being in their senior year [grade 12] (chapter 9).

When looking at the earth's atmosphere, we can focus first on the global circulation of winds. In doing this, the difference between winds that blow above the oceans and winds that blow above the continents should be discussed. In the discussion of areas of high and low pressure, we will deal with the second mode of motion: staunching and loosening. The understanding of the large-scale processes of the weather arises from the penetration of both modes of motion.

It is easier to discuss the structure of ocean currents with their most important aspect, and then proceed with meshing of surface and deep ocean currents. The discussion of the staunching and loosening of water bodies is best done in discussing the tides. When studying the motion of water, it is important to study the precise spatial structure of the ocean basins. In the lithosphere, the third area of the moving earth, we are even further away from direct sensual observation. In using the phenomena of the geology block of grade 9, discussed above, we can demonstrate the motions of the mantle convection and think about their consequences on plate tectonic events. The phenomena of staunching and loosening are shown in the polarity between the events at the mid-ocean ridges and the subduction zones. The formation and back-formation of mountains likewise is closely related to uplifting and descending motions. When we include its layered architecture and function into the discussion of the atmosphere, the correlation to the bodily functions of the human being (which are also a topic within the grade 10 life science block), becomes more obvious. First we could discuss the double function of the multi-layered human skin, which is permeable by certain substances and repellent to other agents. When looking at respiration, we run into the relationship between flowing on the one side and staunching-loosening on the other, with a blending and separation of gaseous and liquid substances.

## *The Flowing Hydrosphere of the Earth*

### *Surface Currents*

Every good atlas contains illustrations of the surface currents of the world oceans, though often omitting illustrations of the currents under the Arctic and Antarctic masses of ice. The three large oceans show at first five partitions of currents (Fig. 3.1). Symmetrically separated by the equatorial region, the Atlantic and Pacific Oceans both contain one current each to run clockwise in the Northern hemisphere and one each running counterclockwise in the Southern hemisphere. Depending on the time of year, the Indian Ocean shows either the same pattern or only just a current in the Southern hemisphere. The strongest current flows from west to east around Antarctica, thereby mixing up all three oceans. Especially remarkable is an east-to-west current below the ice masses of the Antarctic Ocean. The Arctic Ocean shows another kind of behavior: The northern Canada Basin documents a counterclockwise circular current and the northern Greenland Basin a smaller current circling counterclockwise.

Currents covered by ice masses move slower relative to the earth's rotation that is preferably from east to west. Relative to the rotation of the earth, the circumantarctic current flows forward, a phenomenon, which at first sight seems to contradict the principle of inertia. When differentiating currents into so-called cold and warm water currents, then this means relative relations. More accurate would be the following phrasing: One type of current transports warmer water into colder regions and vice versa. The real absolute water temperatures are not involved in this naming convention. It is not fruitful to try to explain the surface currents using the knowledge of the physical laws such as inertia or the Coriolis effect; the encountered inconsistencies are too powerful! It is therefore necessary to move from the surface down into the depth of the oceans. (Garrison 2009; Lalli and Parsons 1997)

*The Ocean Floor and*
*Deep Water Currents*

We come closer to an understanding of the natural laws of currents when we focus our attention to the deep ocean currents. (Fig. 3.2; Blum 1999; Turekian 1985; Turekian 1976; Dietrich and Ulrich 1968) As currents at the ocean floor at an approximate depth of 4000 m occur in a low velocity of only several hundred meters per day, but move a huge water volume, they need to be viewed in connection with the guiding relief of the mid-oceanic ridges. Within the western Atlantic, there is one current flowing from the southern tip of Greenland and another flowing from the Antarctic Weddell Sea towards the Equator. Another stream flows from the Weddell Sea along the coast of Antarctica, through a passageway of the southern Atlantic-Indian Ocean ridge east of Africa to the Gulf of Aden. The western Pacific has a similar situation: A weaker

north-south current originates at Kamchatka, joins a strong current east of Japan which originated in the Indian-Antarctic Basin, broke through the southern Pacific ridge south of New Zealand and pours itself into the northeastern Pacific Basin east of New Zealand and the Society Islands. How do these currents along the ocean floors develop? Above their places of origin, we find the so-called downwelling regions interface at the fast ice-pack ice. At the air-ice-seawater interface, the water continually cools down towards close to the freezing point of seawater of -1.9°C. As the density anomaly of water does not apply to seawater but only to fresh water, the coldest water always exhibits the greatest density and continually sinks down. An additional density increase may occur during ice formation, as the salt brine seeps out of the ice, thus further enhancing the salinity and increasing the sinking rate of seawater. Therefore, the deepwater formation is particularly intense in regions permanently ice-covered such as the Weddell Sea and the Ross Sea in the Southern Ocean.

Deepwater formation, in particular in the Southern Ocean, is of special significance for life processes in the world oceans. The water in the Southern Ocean is enriched with nutrients much more than anywhere else. Because of the permanent downwelling and the lack of stratification, no phytoplankton blooms can establish. Rather, intense phytoplankton blooms prosper where the nutrient-rich deep water emerges again, in particular at the west coasts of Africa and the Americas, in the Tradewinds regions. Phytoplankton forms the basis for a highly productive community of plankton and fish not existing in the oceans anywhere else. The deepwater formation in the coldest oceanic regions thus builds the nutritional basis for intense life processes in distant and much warmer marine regions. This example illustrates how the global oceanic currents act as a circular system—similar to the human being's blood circulation system's nourishing the various organs—which only as an entity provides a meaningful context and that in particular the coldest Antarctic oceanic regions have a unique significance. (Meinhard Simon, personal communication 2010) This is a crucial phase in the instruction. The students experience that we deal with life on earth and not with a physically conclusive explainable process. It is easy to achieve the

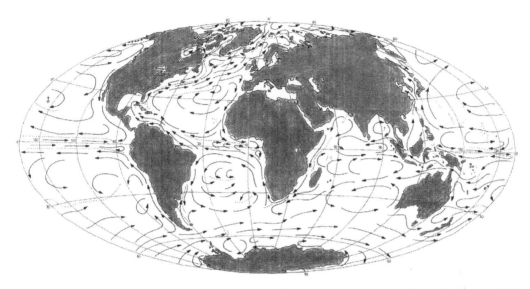

*Fig. 3.1:* Surface Ocean Current Map (redrawn after Marshall and Allen 1979; National Geographic Society 2007; Dietrich and Ulrich 1968)

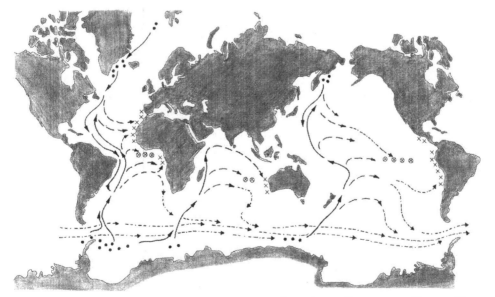

*Fig. 3.2:* Floor- and Deepwater Ocean Current Map. Dots • indicate downwelling sites in cold waters. The water flows along the deepsea floors ➔ and rises along the obstacle of the gently upwards sloping mid-ocean ridges [MOR] to medium sea level I– ➔ . In its further course, the former deepsea water rises [upwelling]: close to continents; in the region of equatorial countercurrents ⊗. (According to Blum 1999; Turekian 1985; Dietrich and Ulrich 1968.)

necessary level of attention when mentioning The Antarctic Treaty of 1959 which came into force on 23 June 1961 and demanded the abandonment of an economical usage of the Antarctica. It was only after the resolution of their convention that more scientific evidence was uncovered to show how sensitive and substantially important this region of the world is for all life on earth.

In 1998, the Protocol on Environmental Protection to the Antarctic Treaty came into force. It bans the exploitation of all

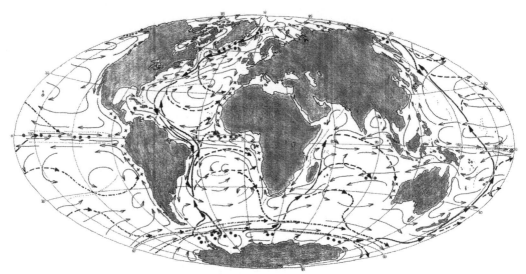

*Fig. 3.3:* Surface and Floor- and Deepwater Ocean Current Map, combined from Figs. 3.1 and 3.2.

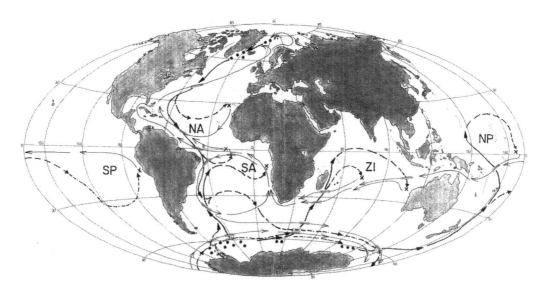

*Fig. 3.4:* The impulses for the motions of the Gulf Stream, originating from four currents forming special lemniscates in the three large oceans. Loops in the northern Atlantic [NA], southern Atlantic [SA], Central Indian Ocean [ZI], northern and southern Pacific [NP and SP]. (Extended after Schmutz 1998)

commercial mineral resources for the next fifty years [until 2048].

## Relationships between Surface and Deepsea Currents

When sea floor currents collide, they escape upwards along the route of least resistance. These are enormous, sluggishly flowing water masses (Figs. 3.2 and 3.3). As there is an insuperable barrier in the west, the steep continental slope, this escape occurs eastwards crosswise to the slowly sloping mid-oceanic ridge. When the current has risen along the ridge to approximately 2000 m depth, its density is slightly dropped due to intermixture, and its water temperature has risen to 1°C. The water body can therefore not sink down into the ocean basin east of the mid-oceanic ridge. Thus the preconditions are in place to enable interactions with the surface currents. Two phenomena occur. As a first result, the so-called equatorial countercurrent develops due to the uprising double drum motion in the equatorial region, which flows opposite to the Tradewinds and currents caused by the Coriolis effect. On the other hand are areas of upwelling nutrient-rich, life-facilitating deepsea water at the western coasts of South and Central America, Africa and Australia. Together with the Tradewinds, these upwelling areas trigger the impulses which keep the surface currents in motion. Without this relationship between deepsea and surface currents, the surface currents would have stopped long ago! It is recommended to prepare exercises for the students to allow them to deepen their work using their self-drawn current maps. When we assume that a colored water body could be observed on its flow through the world's oceans, we can ask the following questions: 1. Which is the most probable way that a water body would flow starting from the coast of Alaska and ending up in the North Sea? 2. Which is the fastest way to get from the Alaskan coast to the North Sea? Choosing almost any starting and target site, the students' answers are interesting as there are no unambiguous standardized answers, but always various alternatives. It is important that the students draw and explain the course while differentiating surface and deep currents.

## The Driving Pulses of the Gulf Stream

The currents shown in Figure 3.3 are also shown in a simplified form in Figure 3.4. The Gulf Stream is not limited to the Atlantic Ocean. The upwelling deepsea currents form spatial lemniscates on four sites of the hydrosphere and stimulate the surface currents which are driving the Gulf Stream in the Central Atlantic. The first small loop is located within the North Atlantic [NA]. The second, more important is located within the northern South Atlantic [SA]. The third loop in the western Central Indian Ocean [ZI], the fourth in the western North Pacific [NP] and the fifth in the South Pacific are driving the surface current which, coming from Indonesia across the Indian Ocean and bending around the southern tip of Africa, combines with the north Equatorial Current, thereby giving the Gulf Stream the required motion impulse in the Central Atlantic close to the Equator.

## The Importance of the Gulf Stream

The students are probably already familiar with the importance of the Gulf Stream for the American and European climates. When comparing the arctic climate of southern Greenland with southern Scandinavia, which lies on approximately the same latitude, you may experience how much relatively warm water the Gulf Stream allows to flow towards Scandinavia to achieve a temperature increase of up to 10°C above solid ground. Current research on ice cores and Atlantic sediments from Greenland shows how important the power and spatial expansion of the Gulf Stream was for the sequence of ice and warm ages [glacials and interglacials] during the last ice age cycle. (Adkins et al. 1997; Bond et al. 1997; Lehman 1997) It was surprising for the climatologists to learn that approximately one third of the carbon dioxide emitted into the atmosphere by human civilization is bound by the downwelling zones that impulse the Gulf Stream and is thus kept away from polluting the atmosphere for several centuries. It is only thanks to this fact that the greenhouse effect, caused by the human activity of burning carbon compounds, did not show more serious consequences up to now (comp. chapter 8, Fig. 8.8). However, calculations show that if there is a substantial global temperature rise, the downwelling movements, especially off the coast of east Greenland, could be reduced substantially. (Ganopolski et al. 1998; Stocker and Schmittner 1997) This would strongly diminish the storage of carbon dioxide in the depth of the oceans and increase the greenhouse effect on earth. On the other hand, the northwest-European climate would get drastically cooler due to the reduction of the Gulf Stream. This would cause unforeseeable complications not only for the European but for the global climate. This very realistic scenario is of such dramatic proportions that it is almost unacceptable to confront the souls of 10th graders with it. Perhaps this topic would fit better into grade 11, where it can be incorporated into a block on the economy of energy (chapter 8) and discussed together with alternatives to the modern trend in today's civilization.

It is important, however, for the teacher to know of the significance of the Gulf Stream and its connection with the cultural events for the entire living world. When compared to the human organization, the Gulf Stream functions as the heart of the living earth. It receives its impulses from the currents of all oceans and affects the climate and the life of the entire earth. You can find further reflections on the relationship between the Gulf Stream and human evolution inspired by Steiner in chapter 9 and Schmutz 1998.

## Tides

The alternation between low tide and high tide can be understood as an expression of the staunching and loosening of water masses. (Koehler 1985) When comparing the water level fluctuations of a harbor with the sequence of movements of moon, earth and sun, it is possible to recognize comparable periods of the rhythms (Fig. 3.5). Every 14 to 15 days a storm tide happens. Within this time we can count the alternation of low and high tide either 27 times in some harbors (Immingham) or 13 to 14 times, half of the period (Don-Son). High tide returns either every 12 hours 25 minutes or every 24 hours 50 minutes. In some locations one can observe an overlay of both rhythms (San Francisco and Manila). The doubling or bisection of a rhythm is similarly described in medicine in relation to the pulsation of blood in the human being. For students who think logically there will be some disappointment when they realize that spring tides do not happen during new moon, when moon and sun are both on the same side of the earth. Moon and sun do not just drag the waters of the ocean towards themselves, but only actuate a rhythmic lifting and lowering of the oceanic water masses. The rhythm itself is further modified by the spatial form of the oceans. In Figure 3.6, we look at the timelines of tidal waves [lines of constant tidal phase, or cotidal lines] of the half-day tide in the Atlantic Ocean. (Dietrich and Ulrich 1968)

Where they meet at a common knot point [the amphidromic point] there is no shift in the water level [almost no vertical water movement]. The lines with the time in hours of the incoming tidal waves illustrate how the crest of the tidal wave changes its position circling counterclockwise [on the Northern hemisphere] around the amphidromic point in 12½ hours. The sense of rotation changes in the Southern hemisphere. Consequently, close to the Equator, cotidal lines have to connect from one amphidromic point to the next, as best seen in the Pacific Ocean. When studying Figure 3.6, it is possible to receive the impression that the tidal waves do not only approximately circle around the amphidromic points, but that the tidal wave of one amphidromic point passes its impulse to the tidal wave of a neighboring amphidromic point. This way, a complex pulsation, caused by the shape of the ocean floors spreads from one site to the next. The graphs of a harbor's water level can be understood only when we correlate the cosmic rhythms and the shape of the earth. To do this clearly asks too much of a 10th grader, as she/he will not deal with astronomical rhythms before grade 11 (chapter 7). It makes more sense, therefore, if students would only learn to know the tidal relationships and experience that a simple physical explanation is not sufficient. This awakens an interest for an investigation of the rhythmic relations between events on earth and cosmic motions.

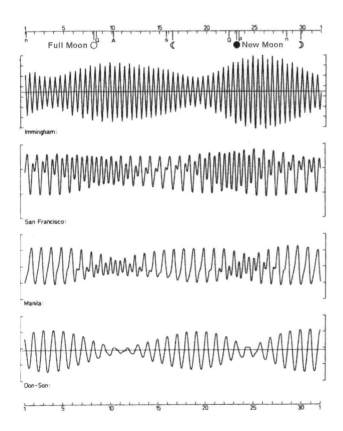

*Fig. 3.5:* Water level graphs of four harbors in March 1936. Immingham at the North Sea shows a half-day rhythm, Don-Son in the Pacific Ocean shows a full-day rhythm. The Pacific harbors San Francisco and Manila document a rhythm overlay. Q = Moon crosses the Equator; A = apogee, and P = perigee of the moon; n = largest northern declination, and s = largest southern declination of the moon. (According to Defant 1973, 1958)

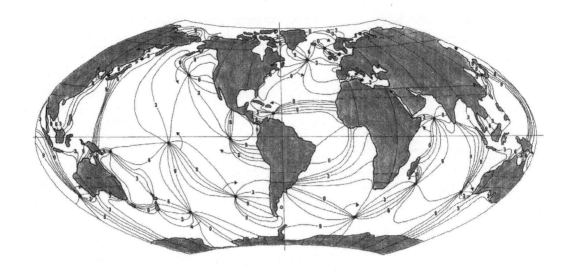

*Fig. 3.6:* Representation of the half-day tidal waves [M2 tidal constituent] in the oceans. The numbers show the incoming times of the corresponding tidal waves in hours. (According to Dietrich and Ulrich 1968)

## Global Wind Circulation

### The Layered Composition [Stratification] of the Atmosphere

As an introduction to the conditions of atmospheric streaming, the teacher can go into the layered composition and by this into the heat and cold coats of the atmosphere. We refrain from describing these layers and the attached phenomena as this material is easily accessible in numerous books (e.g., Battan 1979, 1974; Day and Schaefer 1991; Graedel and Crutzen 1995, 1993; Keidel and Windolf 1986; Rohrbach 1999; Rubin et al. 1984). The student should get a glimpse of how sensitive the dynamic equilibrium [steady state] of the heat and material processes in the various layers is and how important this boundary between the cosmos and the earth is for life on earth.

### Surface Winds

To provide an idea about the method of working together with the students, we look first at the global winds. When it was most of all the relief of the earth's surface which needed to be taken into account when dealing with water currents, it is now the aspect of time that needs to be the center of attention. We will have to distinguish between the motions during different seasons to honor the relation between earth and sun.

The world map of the dominating surface winds in July (Fig. 3.7) allows us to see five large high-pressure areas whirling out above the oceans. In comparison, there is low pressure above the continents, documented by the three inward-whirling low-pressure areas above eastern Canada and southeast Asia and around the margins of Antarctica. The direction of rotation of the whirls is not exactly mirrored at the Equator, but along a line, which shifts between the equatorial regions of the Central Pacific on the Northern hemisphere, first through South America and the Atlantic as much as 10° to 20° Northern latitude and then stretching through Arabia up to 50° Northern latitude above the highlands of Tibet. At this line, the intertropical convergence [ITC], the large whirls of the Northern and Southern hemispheres meet, resulting in a global wind flow from east to west. When observed from outside, both earth poles are characterized by a constant high pressure zone with clockwise outward-whirling winds. In the Southern hemisphere, where there are larger bodies of water, the influence of high-pressure areas is dominant in July and shows three whirls opening up counterclockwise. The Northern hemisphere is dominated by a powerful low pressure area above the large land mass, which includes North Africa, Europe and Asia. The map of surface winds in January (Fig. 3.8) shows most of all large differences on the Northern hemisphere: Centers of clockwise whirling high-pressure are above the land mass of Central Asia and the western side of North America and from southern Europe including the adjacent ocean regions. The two large counterclockwise whirling low-pressure areas are located in the north Pacific and in the north Atlantic above Iceland. On the Southern hemisphere, the three high-pressure areas in July above the southern oceans have moved slightly south. South America, South Africa and northern Australia receive more low-pressure influence. In contrast to July, the intertropical convergence, the region where the northern and southern winds meet has moved on the Southern hemisphere. While it is still oscillating around the Equator above the eastern Pacific and South America, it reaches almost 20° Southern latitude above east Africa, the Indian Ocean, and Australia. Thus we realize that significant seasonal changes are correlated to the large landmasses. Throughout the whole year we find eight large whirls. This is a second indication of the octahedral structure of the earth (see chapters 2 and 9). In July, five oceanic high-pressure whirls coincide with three continental low-pressure whirls. In January, two continental and three oceanic highs compare to three lows at the margins of the oceans.

## Upper Winds

Upper winds, which reach up to the upper boundary of the troposphere and determine the weather, present a much simpler pattern. They flow towards both poles and thus produce high-pressure close to the surface of the poles causing flow-off winds. Above the intertropical convergence [ITC]—which, as is well known, changes its location with the seasons—after they were brought up by a low-pressure area at the ITC, the upper winds flow first towards the poles.

Above the temperate latitudes things are more complicated. Above the near-ground subtropical high-pressure areas, the northerly upper wind meets with a southerly flow, which forms a circulation cell [Ferrel cell] with the near-ground west winds. This drum gives a rhythmically changeable weather to the temperate latitudes (see Fig. 3.9).

When searching—in analogy to the course of flow of the ocean waters—for the impulses of the constant wind motions, you may soon think of the sun, which at noon stands vertically above the Equator. At the summer solstice there is no shadow at noon at the Tropic of Cancer nor around the end of December at the Tropic of Capricorn. But astronomical conditions alone cannot sufficiently explain the course of the winds. Oceanic, and most of all, continental plants play a major role as they regulate the earth's surface temperature, transpiration, and the absorption of the greenhouse gas carbon dioxide.

## Relationships between Wind and Water Currents/Flows

In summary, the global wind systems show symmetry, with the ITC as a seasonally fluctuating plane of symmetry. The surface winds have either a slowing-down or accelerating effect on the surface water currents. Since the shift of the ITC in the Pacific and Atlantic stays above the oceans, there is not much change over the seasons in the pattern of the water currents. In the Indian Ocean, this is completely different. There the ITC changes from a continental orientation in July to an oceanic orientation in January. This explains and differentiates the monsoon phenomenon studied in middle school in a new way into a northeastern monsoon, which influences India and Japan, a northwestern monsoon, which acts in Indonesia, and a pseudo-monsoon which determines the weather in west Australia. The map of the ocean currents in Figure 3.3 shows the situation for the Indian Ocean in January.

## Highs and Lows as Compression and Rarefaction of Air Masses

It is always difficult to satisfactorily explain centers of low- and high-pressure. For instance, a near-ground low is always coupled with a high at the upper limits of the troposphere. A high can be seen as a build-up of air masses, in which the air flows first slowly and then a little faster towards the outside. In contrast, a low always shows a fast spiral-shaped flow into the low-pressure center. The air that flies inside is subsequently rising into the upper troposphere and this way forms the upper high due to a compression effect in the upper troposphere. The low is therefore comparable to a release and getting-into-motion of air masses (Fig. 3.9). The low is also characterized by more dramatic weather around it. The high, which can be described as a phenomenon of congestion, is often preserved for a long time and leads to longer, more stable weather situations.

Flint Beach, Seven Sisters, England. Photo by DSM

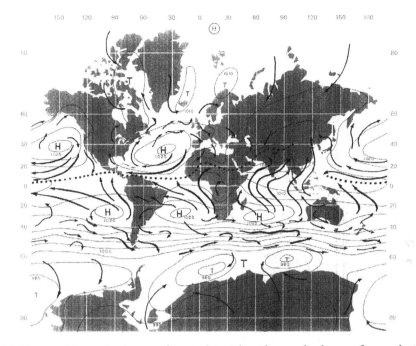

*Fig. 3.7:* World map of the predominant surface winds in July with generalized areas of atmospheric pressure. Short arrows indicate changeable winds, which accompany the traveling cyclones. The dotted line shows the intertropical convergence zone [ITC]. H = High-pressure area; T = Low-pressure area. (From Battan 1979 and Spiess 1993)

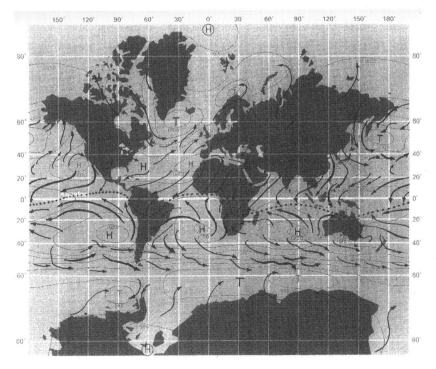

*Fig. 3.8:* World map of the predominant surface winds in January with generalized areas of atmospheric pressure. "T" indicates low pressure areas. Same representation as in Fig. 3.7.

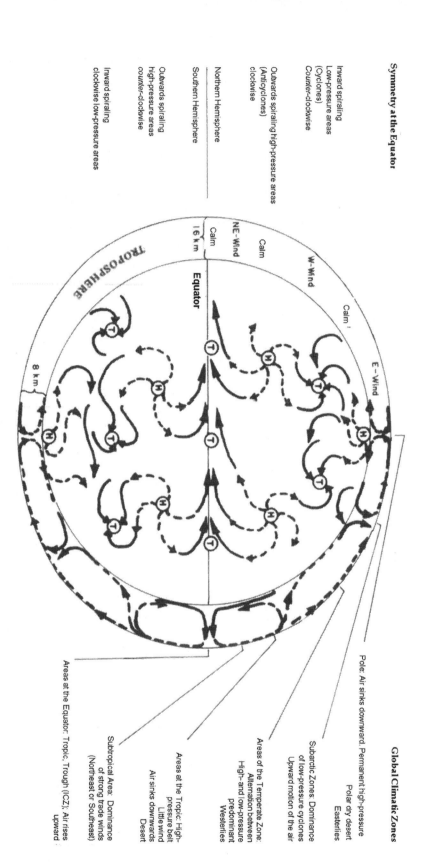

**Symmetry at the Equator**

Inward spiraling
Low-pressure areas
(Cyclones)
Counter-clockwise

Outwards spiraling high-pressure areas
(Anticyclones)
clockwise

Northern Hemisphere

Southern Hemisphere

Outwards spiraling
high-pressure areas
counter-clockwise

Inward spiraling
clockwise low-pressure areas

**Global Climatic Zones**

Pole: Air sinks downward. Permanent high-pressure
Polar dry desert
Easterlies

Subarctic Zones: Dominance
of low-pressure cyclones
Upward motion of the air

Areas of the Temperate Zone:
Alternation between
High- and low-pressure
predominant
Westerlies

Areas at the Tropic: High-
pressure belt
Little wind
Air sinks downwards
Desert

Subtropical Area: Dominance
of strong trade winds
(Northeast or Southeast)

Areas at the Equator: Tropic, Trough (ICZ); Air rises
upward

TROPOSPHERE

16 km

8 km

Calm

NE-Wind

Calm

W-Wind

Equator

E—Wind

*Fig. 3.9:* Schematic representation of the general atmospheric circulation. Interior circle: Face view of the earth with surface winds. Ring: Cross-section through the troposphere, enlarged. (Supplemented after Battan 1979)

58

## Hints about Meteorology

From the careful presentation of the ocean currents (chapter 3), the teacher can develop a global and regional meteorology. I recommend the paperback *Cloud Pictures and Weather Forecast* (Keidel and Windolf 1982) to get more familiar with and understand weather charts. [Battan 1974; Day and Schaefer 1991, and Rubin et al. 1984 are good replacements of this German resource—tr.] It is also helpful to collect weather charts from newspapers and display them during the block. An interpretation of weather charts requires repeated exercise, which could be extended over and above the time course of the block. Long-term recording and interpretation of weather observations can be extended easily to become a research or study project.

## Plate Tectonics as an Expression of Currents within the Earth's Mantle

In chapter 2, I reasoned why a 9th grader cannot already develop a concept for the dynamics of motions as in the case in modern plate tectonics. However, the geology block of grade 9 achieves fundamental preparatory work as it develops the geologic phenomena, which are essential for the understanding of plate tectonics.

After the students experience the global system of air and water currents and start to work into the correlations between the motions with respect to thought, they can go on to think into the third layer of motions that occur within the earth's mantle and crust. Unlike with the oceans and the troposphere, here it is impossible to observe and measure the phenomena first-hand. You are increasingly dependent on seismic processes, your understanding of geological processes and your own mental mobility [kinematical thinking]. This part of the block therefore serves two purposes. On the one side it revisits the subject of the previous year and allows for automatic review. On the other side, it is a matter of applying what was learned in this block in a new area of earth studies. An even further reaching examination of plate tectonics as an expression of a cosmic-earthly relationship structure is not yet developed but will fit into grade 12.

## The Interplay of Expansion Phenomena and Compression Phenomena

One of the most important ideas within the concept of modern plate tectonics is the principle of convection currents inside the earth mantle. Neither the subduction zones nor the mid-ocean ridges [MORs] alone can be made clear. Let us start with the upwelling of mantle material below the MORs. The material that flows upwards from two sides (Fig. 3.10), receives an acceleration in approximately 700 to 50 km depth, resulting in a pressure release. This leads to the partial melting of mantle material. It starts to partially solidify within the last 30–40 km below the earth's surface. This is accompanied by events of tension which periodically dissolve into earthquakes. The spreading at the MOR can be understood only if we add the subduction of oceanic crust under the neighboring continents. Once a subduction zone has been established, the subducing slap starts to "drag" the ocean floor towards the continental margin.

In order to strengthen the idea of mantle convection, which was first postulated in the 1970s, earth scientists produced more and more detailed indications about the interior of the earth by using seismic evaluations of earthquakes. As early as 1989, the geophysicist Vogel postulated that the radius of earth's plastic outer iron-nickel core must be up to ten percent larger on four sites. (Vogel 1993, 1991) He did not grasp the concept of the tetrahedron, but described it indirectly. His

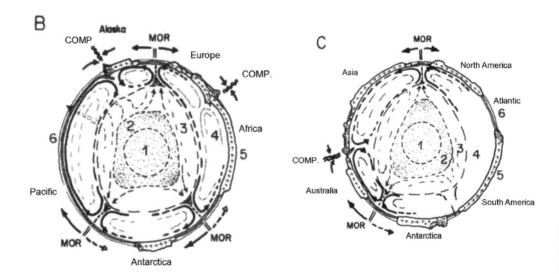

*Fig. 3.10:* Cross-sections of the earth, not drawn to scale; A = Section at about 10° southern latitude, parallel to the Equator. B = North-South section at 20° Eastern longitude. B = North-South section at 80° Western longitude. The triangular margin of the outer core (Vogel 1991, 1993) and the inner mantle (Kellogg et al. 1999) correspond to the section of a tetrahedron with its vertex pointing north. The dilatation tetrahedron shows the same orientation. (Schmutz 1986) Note the Atlantic-Pacific polarity in section A: Due to subduction zones, the convection flow circulates strongly in the Pacific but is congested under Africa. 1 = inner core, 2 = outer core, 3 = inner mantle, 4 = outer mantle, 5 = continental crust, 6 = oceanic crust, MOR = mid-ocean ridge.

idea was controversial for many years, as the predominant models of the earth's interior were revisited.

In 1999, another report was published, in which an extended model of the earth's mantle was proposed. (Kellogg et al. 1999) Due to seismic analyses, the authors postulated a double structure of the earth's mantle. In the upper, 1600 km massive mantle, the circulation of materials takes place, which brings continents via MOR into a horizontal motion. They described the topography of the boundary between the upper and lower mantle in such a way that below the MOR and the dilatation zones, this boundary is up to 400 km closer to the earth's surface. Below the subduction zones, the boundary layer is thought to be dented. As the dilatation zones are situated in tetrahedral arrangement on the globe (Schmutz 1986; comp. Fig. 2.11), the boundary between the upper and lower mantles also shows a tetrahedral shape.

More mobility and a preferred upwelling of mantle material happen at the edges and even more on the corners of this slightly spherical tetrahedron, which lies in the same orientation as the tetrahedron of the outer core. This generates an impulse on the upwelling of more mobile mantle material below the MOR, which explains the maintenance of mantle convection. Conditions for this are the geometric form of the boundaries between different materials in the earth's interior as well as the enhanced chemical and thermal activity along the edges of the tetrahedron.

## Polarity between the Pacific and the Atlantic

A comparison of the water-air and the mantle currents of the two oceans is very informative. Whereas the Equator and the seasonal changing intertropical convergence constitute a plane of symmetry for both the air and water currents, Figure 3.10 shows no equatorial symmetry. The mantle convection therefore does not show a direct relation to the rotation of the earth. Plate tectonic maps show a striking contrast between the Pacific and Atlantic-African spaces. The Pacific region shows a substantial mobility with an active and relatively fast mantle convection, which is especially shown in the large spreading rates of the Pacific ridge. In Africa the contrary is the case. At the end of the grade 9 block, it was shown that the MOR drifts away from Africa almost around the entire continent, without the presence of any compression phenomena in Africa (comp. Fig. 2.12). Below Africa, the mantle is a situation of congestion, also expressed by the abnormal strong heat flow and the large African bulge of the geoid. (Schmutz 1986) The earthquake researcher Pavoni stressed this polarity as early as the 1970s, by observing most of all dilatation earthquakes in the Atlantic and compression earthquakes at the margins of the Pacific. (Pavoni 1991, Pavoni and Muller 2000) The Pacific side shows a circulation which allows crystalline oceanic crust to develop at the MOR and become mobile again at the subduction zones. The Atlantic-African region documents only a ripping open and drifting away of oceanic crust from Africa. This leads to a strong movement of the continents away from Africa. An exception is Eurasia, which shows a compression signature in the Mediterranean and in Iran, which explains the formation of a belt of Alpine folding.

## From the Pulsation of the Single-continent Pangaea to a Multitude of Continents

If you turn the directions of motions shown in Figure 2.11 (chapter 2) around, you get to the paleocontinental world maps, as shown in chapter 9, Figures 9.4 and 9.9 to 9.17. It is striking that the persistence of the deep Pacific Ocean, which changes its relative location on the globe and is wider at the time of the single-continent situation in comparison to a narrowing in the phase of a multitude of continents. The Atlantic, on the other hand, was present in the Ediacaran, and was closed at the beginning of the Paleozoic (Figs. 9.4, 9.9 and 9.10). It only reopened at the beginning of the Mesozoic (Figs. 9.14 to 9.17). In the present South Atlantic, there is already a subduction zone at the Scotia Arc and the Antilles in the Central Atlantic, where the Atlantic is widest.

The map of the ocean floor shows that the Atlantic-oceanic crust is more strongly degraded into the upper mantle, the older it is. It is believed that subduction zones will develop from a certain degradation on, ergo, the Atlantic will be closed again due to the larger spreading dynamics of the Pacific ridge.

Studying the old flattened mountains in Greenland, Scandinavia, Europe, eastern America, Antarctica and Africa suggests that the Atlantic has already opened and closed five times. This is based on the worldwide cycles of orogenesis and subsequent enhanced formation of ocean floor approximately every 350 to 550 million years [MY], known as the Wilson cycle. (Dalziel 1995; Murphy and Nance 1992; Nance et al. 1988)

The face of the earth showed a single continent Pangaea not only 200 to 240 MY ago (Figs. 9.13 and 9.14), but also approximately 750 MY (Fig. 9.4), 1100 MY, 1650 MY, 2100 MY and probably 2600 MY ago, and then for the first time (compare tables in the appendix chapter 9). After the separation, there was a multiple continents stage five times, accompanied by the formation of an "Atlantic Ocean," as described above.

With an eye on the changes in the shape of the earth's face, we are back to the phenomenon of staunching and loosening. This process takes place over very long periods of time. It is remarkable that the alternation between the staunching and loosening and vice versa takes approximately as long as the most dilated known cosmic rhythm, namely the rotation of the Milky Way galaxy around the galactic center over approximately 230 MY.

## Why Not Teach Plate Tectonics before Grade 10?

The internal tracing of the plate tectonic motion dynamics around the globe requires a high degree of spatial imagination and the power of concentration. For example, the spreading at the South Atlantic ridge has effects around the entire globe. The earth does not simply grow through the formation of new oceanic crust. Caused by the western drift of South America, subduction is increased at the Andes subduction zone. On the other hand, Africa is twisted counterclockwise towards Eurasia, which in return causes an east-west transversal motion in Anatolia (compare to the earthquake in northwestern Turkey in August 1999). You will also have to take the southwards drift of Antarctica into account, which causes an increased compression pressure at the Bering Strait.

As you can see, a regional event of motion is directly connected to compression and expansion processes around the globe. And this all happens without the development of either "holes" in the earth crust or massive folds.

To avoid developing plate tectonics into a witchcraft-like show, preliminary exercises are necessary by considering ocean and wind currents first. Ideally, when discussing the elementary plate motions, the teacher can initiate a deeper understanding for the wisdom of the earth's mobility. A complete appreciation of this will not happen before grade 12, when the various placements of oceans and continents at various times are discussed as supporting or inhibiting conditions for the development of life on earth (chapter 9).

## A Look Forward to Additional Subjects

It may seem that the suggested selection of other subjects would be one-sided. There are naturally far more phenomena to be considered which could bring the earth as a whole closer to the student (compare to Rohrbach 1999). A comparison of the earth with other planets and the moon is suggested. A more differentiated atmospheric science with the corresponding light phenomena would be challenging. The pulsating motion of the structure of the earth's magnetic field shows another characteristic of the earth. A global view on ecology could be anticipated in grade 10, if you want to cover other topics in grade 11.

The canon of subjects proposed in this chapter is not mainly focused on the transfer of knowledge [which is not a bad thing], but on the practicing of mental mobility, which is aroused by the grasping of dynamic changes caused by motion processes. By limiting yourself on these subjects, you should be able to facilitate an increased and always self-motivated practicing, and the attempt should be made to internalize elementary aspects of what could be called the life body or etheric body of the earth.

A section of the Grand Canyon in Arizona. Photo by DSM

# Chapter 4

# Crystallography—Tenth Grade

## Preliminary Remarks about the New Block

The traditional block for grade 10 is dedicated to the currents in the air, water and rock spheres of the earth (chapter 3). It is often entitled "The Earth in Its Entirety," "The Earth as an Organism" or "The Earth in Motion." For the preliminary block in grade 6, Steiner demanded a consideration of mineralogy in a geometric way—besides geology. Almost the same phrasing can be found in Steiner's indications for grade 10: "And on the side, do a lot of crystallography, entirely developed out of geometry." (Stockmeyer 1985) The block presented here was created in order to follow this proposal and has been taught over many years with increasing success. It is placed preferably at the beginning of the 10th grade school year.

Three different interwoven aspects provide the characteristic shape of this block. The first task is to observe real crystals and appreciate them in drawing and writing. The second task is the consideration of the science of symmetry, especially mirroring, rotation and central symmetry. Thereby a hard-working practice in the stereometric spatial imagination is needed. Thirdly, the student experiences, how a basic form is modified step by step, resulting in seven crystal systems. These crystal systems describe sufficiently all conformities to natural laws of all naturally-occurring crystals. Besides consequently thinking through the symmetry properties of the various spatial bodies, there is also a lot of drawing in this block, namely of the basic bodies of the crystal systems and their corresponding elements of symmetry.

## Investigations of Rock Crystals

As an introduction to this block, for a couple of days each student could be entrusted with a large rock crystal from a collection of rock crystals as diverse as possible. First, the crystal should be described in detail accompanied by a carefully crafted drawing of a representative view of the crystal. This derives from the achievements of various drawing lessons in the art curriculum; every student may choose his/her preferred way of presentation. The collection of drawings, which should not be done too small, is then lined up in the classroom. The crystals themselves receive a place in an exhibition that is shown throughout the block (Fig. 4.1). In a group discussion it is sorted out which descriptive elements are fitting to all crystals and which to only certain individual or subgroups of crystals. The result has a sobering effect: No natural laws are initially discovered.

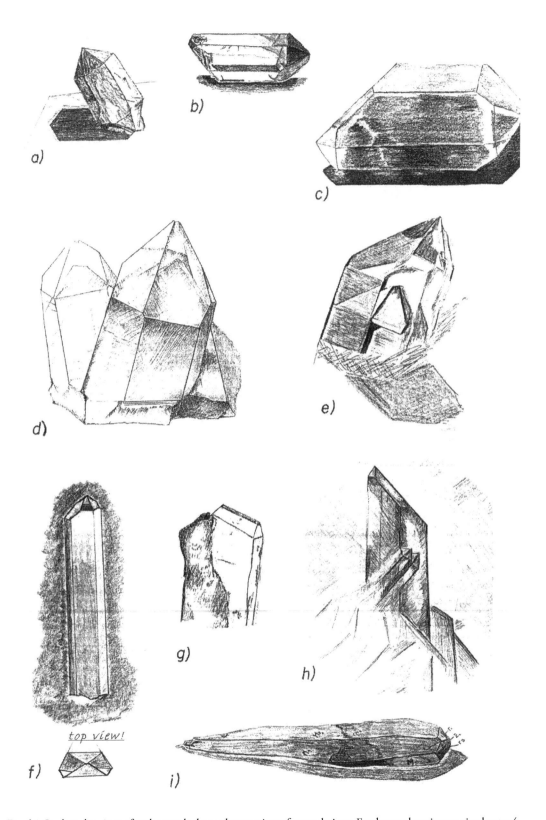

a)

b)

c)

d)

e)

f)

top view!

g)

h)

i)

*Fig. 4.1:* Student drawings of rock crystals show a large variety of area relations. Further explanations are in chapter 4.

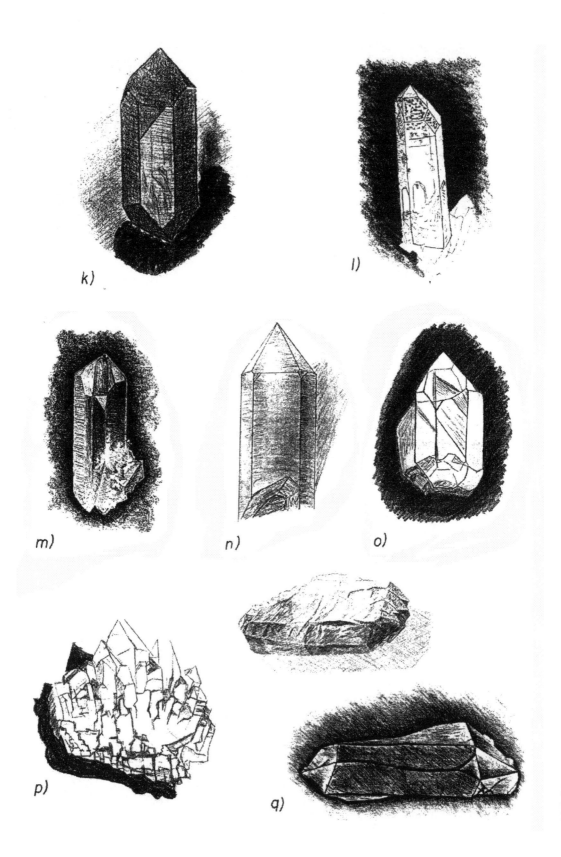

k)

l)

m)

n)

o)

p)

q)

The tips of the pyramids are very different. Not all crystals have even the expected six prismatic surfaces. One specimen shows only five! Some crystals show curved prismatic surfaces. In addition, the roughness and etch traces of the surfaces are very different. But there is another regularity: The non-neighboring prismatic surfaces show a corresponding pattern and the ones in between show another corresponding pattern. The triad becomes increasingly more dominant, and the anticipated hexad disappears. The proportions and numbers of surfaces vary extraordinarily strong. After the assumption occurred, that the angle between the surfaces might be consistent, this needs to be proven by measuring.

### The Law of the Constancy of Interfacial Angles between Corresponding Faces

Each student now measures his/her crystal with a simple do-it-yourself goniometer. This consists of two, nicely trimmed strips of cardboard, which are assembled together on one side with a screw, nut and locknut in a way to allow the opening and shutting of the angle against a small resistance. To read the angle, you need an additional geometry set square or a protractor. Then the different types of faces are named (Fig. 4.2) and the appropriate measuring method discussed. The angle meter must always be in perpendicular position to one of the edges, which the two faces form to be measured. On the next day, a summary table from all measurements is developed on the blackboard—more than 400. The mean of all measurements is computed and an astonishing result emerges. For example, the angle between two neighboring prism faces is always 120°. The law of the constancy of interfacial angles as a hypothesis has been discovered. In addition, also the angles for example between one prismatic face [m] and another neighboring pyramid face [r or z] are by law identical at 142°. Between r and z you are measuring 134° and from r to r or z to z 86°, respectively.

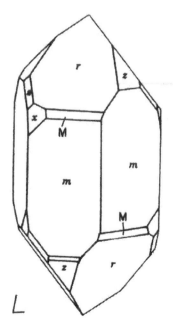

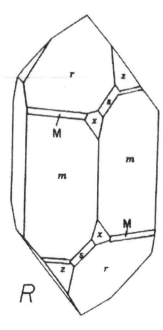

*Fig. 4.2:* Ideal gestalt [shape] of a rock crystal (left) and a rock crystal (right) with the naming of the faces.

Now, the question is: How can this law be proven? Students want to know now. This leads inevitably to the consideration of geometry.

## The Three Important Types of Symmetry

### Mirror Symmetry

First thoughts are about the mirroring of the two-dimensional characteristic—then of the three dimensional characteristics. A body is symmetrically projected onto a mirror plane. Thereby left and right and the direction of rotation of the numbering of the corners of a triangle are exchanged. Above stays above—and the front stays in front (Fig. 4.3). A body contains a mirror plane, if each edge can be brought perpendicular to the mirror plane, and a corresponding perpendicular point on the counter side. This is also valid for planes and edges.

### Rotational Symmetry

Two bodies are in rotational symmetry to each other if the one body can be rotated for a certain value around a rotational axis and can be superimposed with a second object. We realize that the order number of the rotation axis is important. At first, we will distinguish between twofold [order 2 = dyad, 2 x 180° rotation], threefold [order 3 = triad, 3 x 120°], fourfold [order 4 = tetrad, 4 x 90°] and sixfold [order 6 = hexad, 6 x 60°]. According to international notation (Kleber 1965), the order of the rotation axis is marked with a dot for dyadic a triangle for triad, and so forth (e.g., Fig. 4.4). For example, a body contains a twofold axis of rotation if each element [point, line, area] can be rotated by 180° around the rotation axis and can be superimposed to an existing element (Fig. 4.3). Left becomes right, front becomes back, superior stays superior, and the sense of rotation in the triangle enumeration stays the same.

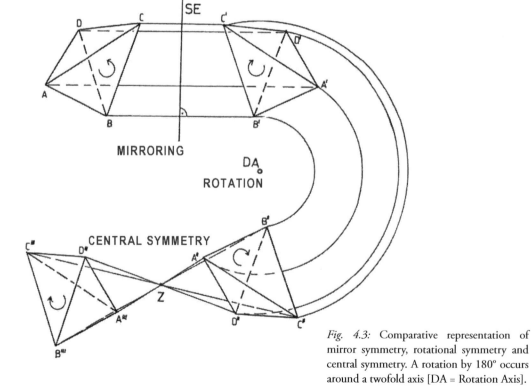

*Fig. 4.3:* Comparative representation of mirror symmetry, rotational symmetry and central symmetry. A rotation by 180° occurs around a twofold axis [DA = Rotation Axis].

### Central Symmetry

When we follow every element of a body to a symmetry center and continue on the opposite side in the same direction and the same length, then these two bodies are centrally symmetric or point symmetric. Now everything is exchanged: Left becomes right, front becomes back, and upside becomes downside. The sense of rotation in the triangle enumeration changes (Fig. 4.3). A body contains a symmetry center if each element [point, line, area] is led through the center and continues onto the opposite side in the same direction and for the same length and can then be superimposed to the opposite element.

It will now become clear that, in bodies which contain one or more of these symmetries, the same angles must show up between corresponding faces. The task is now to investigate the symmetry of spatial bodies.

### Symmetry Elements of the Tetrahedron

The most basic spatial body is the tetrahedron, which is formed by the minimum of four planes heading for each other. The regular tetrahedron, consisting out of four identical, equilateral triangles, develops if the planes are in balanced angular relations to each other. When searching for mirror planes in this simplest of the five platonic bodies, you will soon discover something: One plane, defined by one of the edges of the tetrahedron and the middle of the perpendicular opposing edge, divides the tetrahedron into two mirror identical parts. As such a plane can be defined for each of the tetrahedron's edges; the tetrahedron contains six mirror planes. These planes all meet in the symmetry center of the tetrahedron and intersect in two each along three intersection lines, which each marks the connection between the middle of the tetrahedron's

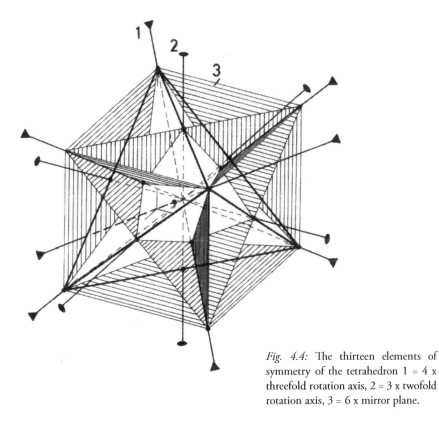

*Fig. 4.4:* The thirteen elements of symmetry of the tetrahedron 1 = 4 x threefold rotation axis, 2 = 3 x twofold rotation axis, 3 = 6 x mirror plane.

edges. Additional intersection lines are found in quadruple copies, in which three mirror edges meet. These four lines develop from the arc between one corner of the tetrahedron and the middle of the surface of the opposite triangular face. Rotation symmetries can be discovered when you imagine the body would be pierced by axes, which function as two-, three-, four- or sixfold rotation axes. In looking at one corner of the tetrahedron, a threefold rotation axis is discovered, which runs from one corner to the opposite surface center. Because I can do this for all four corners, I find four threefold rotation axes, which all meet in the center point of the tetrahedron. When looking on one edge of the tetrahedron, I find an additional rotation axis. This one runs from the middle of the edge to the middle of the perpendicularly opposite edge and is twofold, superimposing the tetrahedron after a rotation by 180°. Because I need two of the six edges for fixing this kind of rotation axis, there are three twofold rotation axes, which also all meet in the center point of the tetrahedron.

It is noticeable that the rotation axes we found are identical with the two kinds of intersection lines of the six mirror planes described above.

You will look in vain for a central- or point symmetry because it is the nature of the tetrahedron to have a different opposite. Opposite of an edge [point] is a triangular face [plane], and opposite of an edge is another edge twisted by 90°.

Figure 4.4 shows all of the thirteen discovered elements of symmetry, drawn in relation to the tetrahedron, to show the mirror planes between the tetrahedron and the enveloping cube [drawing cube].

The rotation axes with the symbols for the order are shown outside of the tetrahedron in bold and within the tetrahedron as dashed. The tetrahedron itself is shown as if the mirror planes were transparent, with something like an interior illumination. These specifics are then described in more detail so the students can copy this representation accurately into their main lesson book—a challenging practice in spatial imagination. The following drawings of spatial bodies and the appropriate symmetries also should be drawn into their books. The so-called drawing cube functions as a pattern, into which all spatial bodies can be inscribed. The students will draw this cube once on a extra sheet. From this drawing, the corners or other special points are punched onto the sheet for the new drawing, resulting in a reappearance of the identical cube in all drawings. The cube is developed by drawing a real cube carefully in parallel perspective and at such a viewing angle to avoid disturbing overlaps of corners, diagonals or edges. The students will be amazed that the resulting cube will look cubic only when the drawing paper is viewed along a certain line of view and appears distorted in width or height when viewed from another perspective.

## The Kepler Star [Kepler-Poinsot Polyhedron] or the Polarity of the Tetrahedron

With an attentive 10th grade class it is highly valuable to do this exercise in imagination, showing the polarity of the tetrahedron. Try to imagine an eternal, wide and nondifferentiated space. From the bottom, a plane comes into view separating two areas, one below and one above the plane. A second plane comes into view diagonally from the top and from behind. Now, there are four areas. The new discovery is an intersection line, running horizontally left-right in front of me, defining the relation between both planes. From this intersection line, I can expand into four wedge-formed areas. Now, another plane approaches, coming closer quickly from the right in front of me. When I bend down onto the previously discovered intersection line, I realize that it is interrupted at one place, namely where two additional intersection lines have come into the picture and which form a pointy corner. When I now investigate the existing areas, I discover eight of them, which radiate from this edge. When I pay attention to the fourth plane, which floats closer from the top in front and left side, I discover an entirely new world: A closed spatial body has developed. I recognize a new corner and two additional intersection lines in the relation between planes 1, 2 and 4. The relation between the planes 1, 3 and 4 shows me another corner and a new intersection line.

The relationship between planes 2, 3 and 4 results in the fourth corner. Therefore I can now envision a spatial structure consisting of four corners, six edges as sections of the six intersection lines and four triangular surfaces as interior boundaries of the four moved planes: the tetrahedron, the most basic spatial body.

Now I move the four planes evenly towards each other. The tetrahedron becomes smaller and smaller, until it is shrunk to a single point. Now it exists only in my thoughts. If I continue to move the four planes, a new, constantly growing tetrahedron emerges. Compared to the first tetrahedron, it is turned around in terms of front/back, bottom/top and left/right. A tetrahedron, polar to the first, has come into being [polar tetrahedron]. If I let both spatial bodies, imagined in the same size, intersect each other by sharing the same center of gravity, a starry structure with eight tips results, the so-called "Kepler Star" [Kepler polyhedron] (Fig. 4.5). The spatial structure resulting from the intersection of both tetrahedrons is another platonic solid, an octahedron.

The envelope of the Kepler Star, which is the arc between the eight corners, results

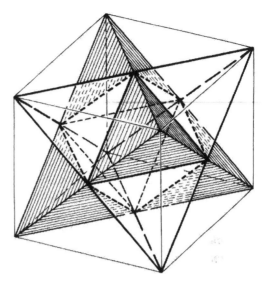

*Fig. 4.5:* Two polar tetrahedrons with the octahedral core and the cube envelope

in a cube, which is another platonic solid. Envelope and core of the Kepler Star are therefore cube and octahedron, respectively.

## The Symmetry Elements of Octahedron, Cube and Kepler Star

When one looks at the octahedron, the assumption arises at once that it contains a higher and more complex symmetry than the tetrahedron. One plane, defined by one of the edges of the tetrahedron and the middle of the perpendicular opposing edge, divides the tetrahedron into two mirror identical parts. There are three mirror planes of this kind. A new kind of mirror plane comes into being when the plane is intersecting perpendicularly through one edge and the parallel opposite edge along the middle of the edges. As two of the twelve existing edges are used for each mirror plane, there are six mirror planes of this kind. The octahedron features central symmetry, as for each element the body has a corresponding opposite. The threefold rotation symmetry known about the tetrahedron is found when piercing the rotation axis from the center of a triangular surface to the center of the opposite surface.

Again, the number of threefold rotation axes is quickly found: From the eight existing triangular surfaces, one pair is needed for each rotation axis. There are therefore four threefold rotation axes. When looking at one corner of the octahedron, one finds one fourfold rotation axis: The body therefore superimposes congruently four times after a rotation by 90° each. There are three such rotation axes since the six corners divide into three pairs. The twofold rotation axis runs from one center of the edge to the center of the edge on the opposite parallel edge.

The body superimposes congruently twice after 180° rotation each. There are six twofold rotation axes, since two edges each are needed from the twelve edges to define a rotation axis. The entirety of all the symmetry elements of a regular octahedron is shown in

Figure 4.6. The nine mirror planes intersect in the central symmetry point. The three threefold rotation axes run along the place where also the three mirror planes of the one kind intersect. There, where the three mirror planes of the second kind are intersecting in pairs each, we find the fourfold rotation axes. The twofold rotation axes are located at the intersection lines of each mirror plane of the first kind with a mirror plane of the second kind. From this finding, the student realizes the connection between rotation axis, mirror plane and symmetry center.

After studying and drawing the symmetry characteristics of the octahedron (Fig. 4.6), you can ask the students to do the same for the cube and the Kepler Star as homework. The results are very puzzling for them: All three bodies, the Kepler Star, as

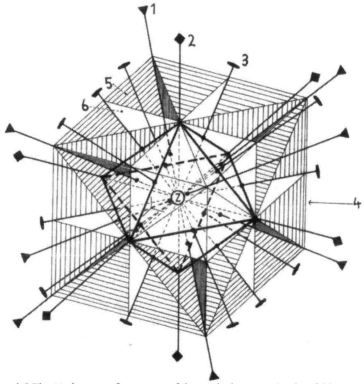

*Fig. 4.6:* The 23 elements of symmetry of the octahedron. 1 = 4 x threefold rotation axes, 2 = 3 x fourfold rotation axes, 3 = 6 x twofold rotation axes, 4 = symmetry center, 5 = 6 x mirror plane of the first kind, 6 = 3 x mirror edges of the second kind, Z = symmetry center

well as its octahedral core and cube envelope, contain the same symmetry elements and, compared to the tetrahedron, more complexity. There are two kinds of mirror planes and three kinds of rotation axes. In addition, there is a symmetry center. In both bodies, tetrahedron and octahedron, there are four threefold rotation axes. At the place, where in the tetrahedron are the three twofold rotation axes, you find in the cube the three fourfold rotation axes and mirror planes of the tetrahedron. The mirror planes of the tetrahedron can also be found in the octahedron.

You can start in two ways when drawing a tetrahedron into a drawing cube, with one corner of the tetrahedron below either left or right. As there is no symmetry center, you can talk about a left form or right form of the tetrahedron. When the left- and right forms intersect each other, the Kepler Star with its higher symmetry emerges.

## The Development of the Seven Crystal Systems from the Archetype of the Tetrahedron

When the students have worked well into these themes, which can easily take ten days, you can now start to develop the schematic of the seven crystal systems in a stepwise metamorphosis of the tetrahedron. (Schmutz 1986) The path is clear: grasping of the basic form and explanation of the symmetry; changing the form and stopping to verify the symmetry again; modifying the form again; stopping and clarifying the symmetry, and so on. For each system, you should add, draw and clarify appropriate examples of crystals. In this block, drawing is as important as logic thinking.

### The Cubic System

The basics for an understanding of the cubic system were presented during the discussion of the tetrahedron (Fig. 4.4) and octahedron (Fig. 4.6). In nature, you can find a large variety of crystals to show the common feature of four threefold rotation axes. It is not necessary for the work in grade 10 to go systematically through all thirty-two crystal classes that are contained in seven crystal systems. (Chatterjee 2008; De Graef and McHenry 2007; Kleber et al. 1990) If appropriate, a system is separated into two groups. This is also the case here: The forms that contain only four times a threefold rotation axis are counted as parts of the cubic-tetrahedral subsystem (2 classes). Crystals that show three times a fourfolded rotation axis belong to the cubic-octahedral subsystem (3 classes). Examples for the cubic-tetrahedral system are the rare minerals tetraedrite, zinc blende in tetrahedral form and the common mineral pyrite, an iron sulfur compound. (There is a closer inspection of pyrite later in this chapter.) In the cubic-octahedral system, the so-called noble materials such as the metals gold, silver, copper, lead, iron and platinum, and the metal ores, lead glance and magnetite, crystallize mostly in octahedral, cubic, or rhombic dodecahedral form. Also the more complex group of garnets, fluorite, rock salt and, last but not least, the octahedral diamond as the purest carbon belong to this subgroup.

Rewarding objects for the drawing and investigation are garnets [rhombic dodecahedra or icosatetrahedra], magnetite [combination of octahedra and cubes], rock salt [cubes] and naturally pyrites (see pyrite, later in this chapter). Specialties are sodium chlorate minerals, which show left- and right-forms due to the lack of a symmetry center.

### The Tetragonal System

The starting point for the development of the tetragonal system is the cube or the octahedron. If you elongate or compress an octahedron along one of its fourfold rotation axis, you receive a double pyramid with a square base (Fig. 4.7a). When investigating all symmetry elements of this body, you get to the following result: The fourfold rotation axis, which serves as a tensile axis, is maintained. Both additional, perpendicularly positioned rotation axes are reduced to the twofold order. The four threefold rotation axes are not valid anymore because the triangular surfaces are only isosceles and no longer equilateral, as was the case in the octahedron.

Of the six twofold rotation axes of the octahedron, those two are preserved that stretch from the middle of the edge to the opposite middle of the edge of the base square; the remaining four are dropped— for the same reasons as in the case of the threefold axes. The symmetry center remains, as the opposite to every element is also preserved. Mirror planes are also the planes of the base square and the four planes that intersect the remaining fourfold rotation axis. The result can be schematically drawn like this: The newly created tetragonal system is characterized by one and only one fourfold rotation axis.

## The Orthorhombic System

When we stop after the first metamorphosis of the octahedron and explain the tetragonal system, now we modify the stretched cube to the next level. The twofold rotation axis runs from one center of the edge to the center of the edge on the opposite parallel edge. This results in a cuboid, consisting of three different pairs of rectangles (Fig. 4.8). Again we begin with the precursor, the tetragonal system, and investigate which symmetry elements are still there and show them schematically:

The higher-fold rotation axis turns into a twofold because the perpendicular cuboid surface is now not a square but a rectangle. The two mirror planes and the two threefold rotation axes, which were running diagonally inside the tetragonal system, have to disappear as well, because we do not have any diagonals of a square anymore but only rectangular diagonals. The orthorhombic system is therefore characterized by three twofold rotation axes that are perpendicular to each other. The mineral bitter salt is an example, in which only the rotation axes are present. The minerals aragonite, sulfur, anhydrite, baryte, olivine and topas contain all possible

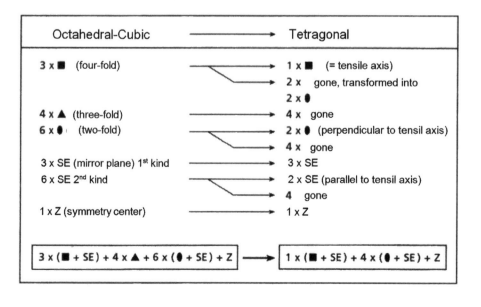

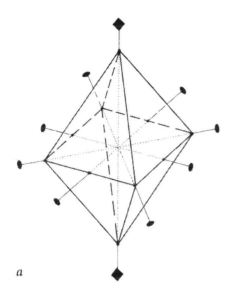

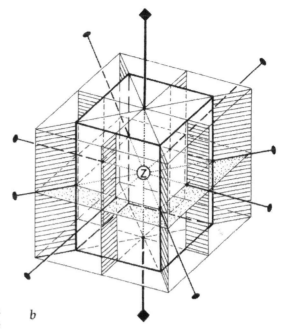

*a*

*b*

*Fig. 4.7*: a) The tetragonal system derived from a stretched octahedron with its rotation axes. b) The entire eleven symmetry elements of the tetragonal system with the basic form of a rectangle with a square base. They can be joined by the symmetry center, a maximum of five mirror planes and four twofold rotation axes (seven classes altogether). The following better-known minerals crystallize in the tetragonal system: chalcopyrite with two twofold rotation axes and two mirror planes, as well as tin stone, rutile and zircon. The last three minerals contain all symmetry elements shown in the above schema. Zircons are compressed octahedra, and rutile contains the tetragonal prism. b) shows a stretched cube within the drawing cube to show the sum of all symmetry elements.

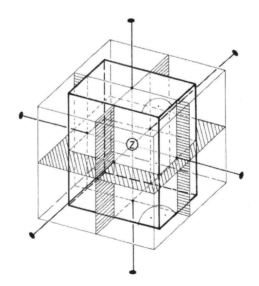

*Fig. 4.8:* Complete symmetry elements of the orthorhombic system with a general cuboid as its basic form

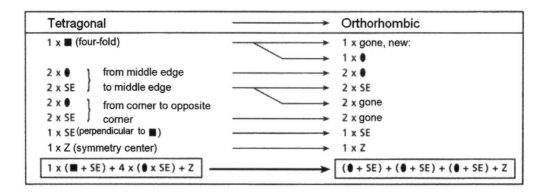

| Tetragonal | | → | Orthorhombic |
|---|---|---|---|
| 1 x ■ (four-fold) | | → | 1 x gone, new: |
| | | → | 1 x ● |
| 2 x ● } from middle edge | | → | 2 x ● |
| 2 x SE } to middle edge | | → | 2 x SE |
| 2 x ● } from corner to opposite | | → | 2 x gone |
| 2 x SE } corner | | → | 2 x gone |
| 1 x SE (perpendicular to ■) | | → | 1 x SE |
| 1 x Z (symmetry center) | | → | 1 x Z |

1 x (■ + SE) + 4 x (● x SE) + Z ⟶ (● + SE) + (● + SE) + (● + SE) + Z

orthorhombic elements of symmetry. The mineral baryte is especially suited to illustrate this crystal system, as the rotation axes are easy to find.

### *The Monoclinic System*

When trying to further compress or stretch the general cuboid of the orthorhombic system along the twofold rotation axes, one is left still with a general cuboid. A new metamorphosis will occur when imagining the cuboid as a wire model. Place the wire cuboid on one if its edges and drag or push on the diagonally opposite edge. This results in a once slantingly pressed cuboid, which consists of two pairs of cuboids and a oblique-angled pair of parallelograms (Fig. 4.9). Again, we investigate the symmetry of the

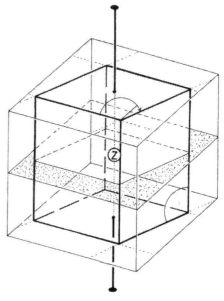

*Fig. 4.9:* The maximum of three symmetry elements of the monoclinic systems, drawn into the basic form of a once unevenly-compressed cuboid.

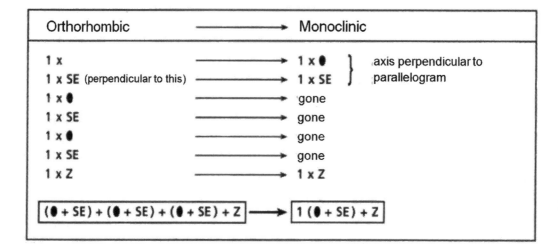

| Orthorhombic | | → | Monoclinic | |
|---|---|---|---|---|
| 1 x | | → | 1 x ● } | axis perpendicular to |
| 1 x SE (perpendicular to this) | | → | 1 x SE } | parallelogram |
| 1 x ● | | → | gone | |
| 1 x SE | | → | gone | |
| 1 x ● | | → | gone | |
| 1 x SE | | → | gone | |
| 1 x Z | | → | 1 x Z | |

(● + SE) + (● + SE) + (● + SE) + Z ⟶ 1 (● + SE) + Z

resulting body by comparing it to the former body. The schema becomes even simpler: Only the one axis is maintained that sits perpendicularly to the pair of parallelograms, together with the corresponding, mirror plane, which is parallel to the parallelogram.

There is no symmetry center as rotation axes are accompanied by perpendicular mirror planes. The orthorhombic rotation axes, which are now parallel to the parallelogram, vanish. Typical of the monoclinic system is that it contains one and only one twofold rotation axis and/or one mirror plane. These include the common minerals such as the various types of mica, the group of hornblende, serpentine, potash feldspar [as mixed crystal twinned in triclinic forms], and gypsum as well as the copper ores malachite and azurite. As an extra touch, the substances tartaric acid, lactose and saccharine crystallize as well in a left as a right form because they possess only a twofold rotation axis and by this no symmetry center. The ideal gestalt of the once unevenly compressed cuboid with a maximum of monoclinic symmetry elements is shown in Figure 4.9.

## The Triclinic System

When unevenly compressing the above mentioned monoclinic body at an additional pair of edges, even the last remaining rotation axis disappears, and we find only bodies without any symmetry or ones with just the remaining symmetry center. We can call the partial system without symmetry triclinic asymmetrical and the partial system with just central symmetry triclinic central symmetric. Figure 4.10 shows the double unevenly compressed cuboid, which consists only of pairs of parallelograms.

The hugely abandoned feldspars in the triclinic central symmetric system crystallize, often twinning up with monoclinic

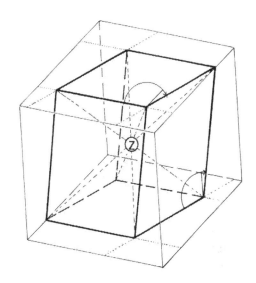

*Fig. 4.10:* The only remaining symmetry center of the triclinic system, drawn into the basic shape of the double unevenly-compressed common cuboid.

crystals. Albite, pure sodium feldspar, crystallizes triclinic. Another example is the clinopyroxenes. Only organic compounds such as calcium thiosulphate crystallize in the asymmetrical partial system. Crystals of copper vitriol [copper sulphate], which students like to grow, contain a symmetry center. This means that for every surface, a parallel surface on the opposite side can be found.

## The Trigonal System

After trying all varieties of metamorphosis of the octahedron even into asymmetry, we have to go all the way back to the archetype of the tetrahedron. The tetrahedron can also be stretched or compressed at one of the threefold rotation axes. This results in a threefold pyramid (Fig. 4.11). Using a schema, we can show again how the metamorphosis of the cubic-tetrahedral system leads to the trigonal system:

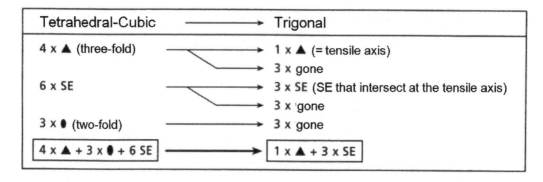

| Tetrahedral-Cubic | → | Trigonal |
|---|---|---|
| 4 x ▲ (three-fold) | → | 1 x ▲ (= tensile axis) |
| | → | 3 x gone |
| 6 x SE | → | 3 x SE (SE that intersect at the tensile axis) |
| | → | 3 x gone |
| 3 x ◗ (two-fold) | → | 3 x gone |
| 4 x ▲ + 3 x ◗ + 6 SE | → | 1 x ▲ + 3 x SE |

The axis of stretching is maintained as a threefold rotation axis, as we already found in the corresponding tetragonal system. The remaining three copies are discontinued, as the three pyramid faces are just isosceles. Likewise, the twofold rotation axes have to vanish because two opposite edges are no longer perpendicular to each other. The three mirror planes that are in a skewed angle to the remaining rotation axis do not exist anymore (Fig. 4.11). The determining character of the trigonal systems is the presence of one and only one threefold rotation axis. There are five classes within this system. We distinguish between the trigonal system with a low symmetry and the trigonal system with a high symmetry. The first contains only ordinary rotation axes and can develop into left [laevo] and right [dextro] forms. The important quartz or rock crystal and cinnabar belong to this group. Tourmaline, dolomite, calcite and corundum crystallize in the second partial system.

Let us take quartz as example of the trigonal low symmetric system: The basic form, which results from the relation between the threefold rotation axis and three perpendicular twofold rotation axes, is the trigonal trapezohedron (Fig. 4.12). It is placed into the drawing cube in such a way that the threefold rotation axes are perpendicular to the center, and one of the threefold rotation axes is perpendicular to one of the faces of the cube.

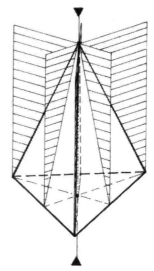

*Fig. 4.11:* Threefold pyramid, developed by stretching a regular tetrahedron at a threefold rotation axis.

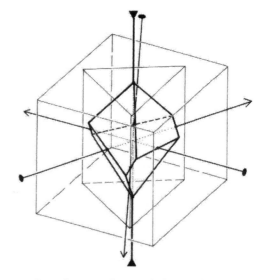

*Fig. 4.12:* The trigonal trapezohedron as the common basic form of quartz. The three twofold rotation axes are perpendicular to the threefold main axis.

The upward-pointing pyramid is rotated by 60° against the lower pyramid. This leads to the possibility of laevo and dextro forms, which is the case in quartz. Additional aspects of quartz are discussed in this chapter.

Take calcite as example of the trigonal high symmetric system: This adds three mirror planes and a symmetry center to the threefold and the three twofold axes. Figure 4.13 shows the basic form of a double pyramid with a bordering crown [trigonal scalenohedron]. Calcite crystallizes in the trigonal class with the highest symmetry.

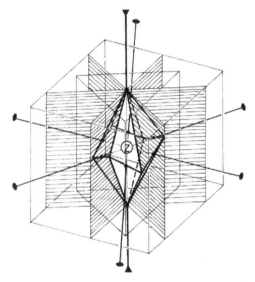

*Fig. 4.13:* The symmetry elements of calcites with the basic form of a trigonal scalenohedron.

### The Hexagonal System

If we now, for example, combine the laevo and dextro forms of the low trigonal system, in the same way we did before for the left- and right-forms of the tetrahedron in order to get to the octahedron, we get to a body with the symmetry of the hexagonal system. The threefold rotation axis is increased to a sixfold, causing the body to superimpose congruently six times after a rotation of 60° each. The complete system also contains three twofold rotation axes together with mirror planes of the first kind, and the same [second kind] but rotated 60°. The ensemble of symmetry is completed by a mirror plane, which is perpendicular to the sixfold rotation axis and the symmetry center (Fig. 4.14). The basic form is the dihexagonal dipyramid, a body constructed of 24 identical isosceles triangles. To simplify things, Figure 4.14 shows the six-sided prism.

Only rather rare minerals crystallize in the hexagonal system. Examples are a side form of feldspars, nepheline, calcium phosphates apatite and beryl, and a special crystallization form of carbon, graphite. Common for all hexagonal crystals is the presence of one and only one sixfold rotation axis.

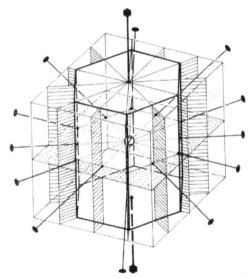

*Fig. 4.14:* All possible elements of symmetry in the hexagonal system drawn into the basic form of a hexagonal prism

### Summary

In this walkthrough of the metamorphosis of crystals, we can show how all possible crystal systems can be geometrically derived from the basic form of the tetrahedron in such a way that allows the 10th graders to follow. Connections are established by the stepwise, intellectual discussion of the possibilities of the tetrahedron. This is an example of an elementary thought metamorphosis that can be grasped through geometry. This highly-appropriate exercise for grade 10 is an important preliminary exercise to prepare for the mental experience of other metamorphoses, e.g., of the plant, which are introduced at the end of high school. The aim is to develop a kind of thinking that is not only imitation/emulation but contains a high portion of self-design that is created by the student him/herself. It is beneficial for the students of grade 10 to stop after each section and note the intermediate results in a drawing. Another experience will be added towards the end of the block: The realm of crystals can be understood as a complete, inherent logical system which does not develop in a series of facts just strung together like the elements of symmetry, but from an evolutionary process. Ordinary textbook crystallography advances cumulatively: One symmetry element is added to the next. In this example it should become clear again, how the pedagogic trick of moving from the whole to its parts can lead to a spiritual economy. Students understand the connection between the seven crystal systems as it is schematically shown in Figures 4.15a and b. This has been verified repeatedly in the final test results of students.

*Fig. 4.15a:* The seven crystal systems developed from the archetype of a regular tetrahedron in nine steps. 1 = Tetrahedron. 2 = Joining left- and right-forms leads to the octahedron and hexahedron, respectively. 3 = Stretching at four edges of the cube results in a cuboid with a square base. 4 = Metamorphosis of the base square to a rectangle [stretching] results in a basic cuboid. 5 = Uneven compression on the cuboid results in the beginnings of a rhomboid [compression]. 6 = The next step of compression leads to the rhombohedron. 7 = Dragging on one tip produces a form without symmetry. 8 = One tip of the tetrahedron is dragged outwards. 9 = The connection of left- and right-forms produces the sixfold pyramid.

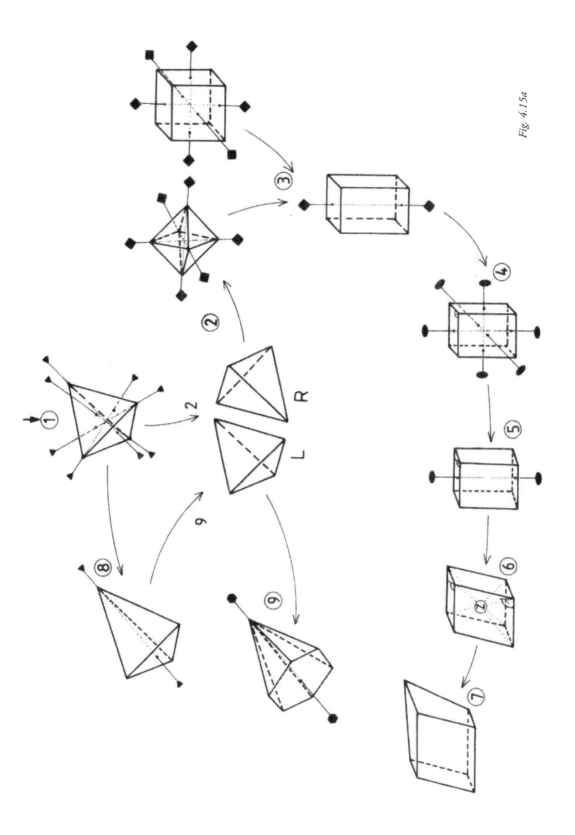

*Fig. 4.15a*

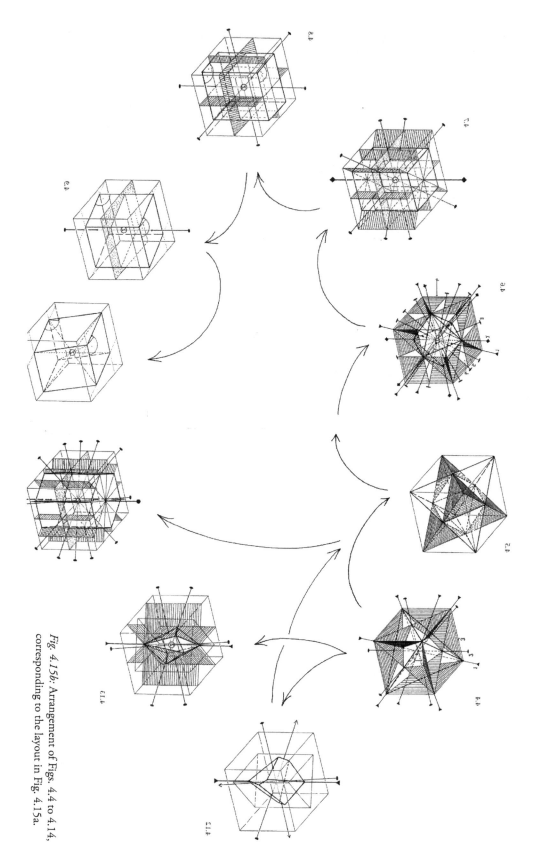

Fig. 4.15b: Arrangement of Figs. 4.4 to 4.14, corresponding to the layout in Fig. 4.15a.

## The Creation of a Crystal Form Derived from the Law of Symmetry

### *Garnet as a Rhombic Dodecahedron*

After it becomes clear that a certain substance is always formed according to the same symmetry law, you can continue to the next step: the creation of a crystal form derived from a law of symmetry. Let us look at an example: The cubic-octahedral law with three fourfold and four threefold rotation axes is given. Position these axes within the drawing cube according to the pattern shown in Figure 4.6. Now try to imagine that one plane comes from the outside and positions itself to the drawing cube in such a way that it cuts out a wedge from it (see plane k1 in Fig. 4.16a). Now rotate this plane k1 according to the given rotation axes and draw the result. At first it is advisable to rotate around the perpendicularly-oriented fourfold rotation axis. After the four steps, a four-sided pyramid develops on top of the drawing cube with its tip called B. As the next step, rotate the plane k1 around a threefold rotation axis which sticks out of the right front top of the drawing cube. This produces the cutting planes k5 and k6, and k6 is identical to k2. Now intersect k1 with k5, v with k6, and k5 with k6, each. The three intersection lines s15, s16 and s56 intersect at point A, which lies on the threefold rotation axis selected for the rotation. You may continue in this way by first rotating the gained intersects k5 and k6 around the next threefold rotation axis. You can also rotate around the two remaining fourfold rotation axes and draw the resulting new intersection lines between the k-planes. The intersection lines so gained are the edges of a new body consisting of twelve rhombi: A regular rhombic dodecahedron is created.

It is possible to accelerate the construction process by rotating the constructed point A, which is one of the edges of the rhombic dodecahedron, around the three fourfold rotation axes. In this way you quickly gain eight corners of the studied body. These eight corners have to lie on the threefold rotation axes [A1 to A8]. In this way you have found another simplification for the drawing. Now do the same thing for tip B. This point is also rotated around the three fourfold rotation axes and results in the finding of B1 to B6. When you connect one B-point with the corresponding neighboring A-point, you get to the twenty-four edges of the rhombic dodecahedron. The body gained by this contains six edges where four rhombi each meet and eight edges where three rhombi get together. In each case, the lengths of the edges and rhombi are the same. When looking at the world of crystals, you discover, e.g., the mineral garnet that forms this gestalt [shape].

A crystal develops therefore out of the relationship of the crystal system with selected planes. The characteristic of these planes is that they cut the so-called main axes [in our case, the three fourfold rotation axes] in a simple number ratio. 1: ∞. For people, who already know crystallography, it should be noted that we are referring to the law of the Miller Indices. This law says that each type of crystal shows a specific crystallographic axial ratio. This ratio is always expressed in small natural numbers. The indices are the reciprocal values of the axial ratio and are named with the three letters hkl. When written in parentheses, the indices for the rhombic dodecahedron are therefore [110]. When using the indices [100] for the octahedral-cubic system, the result of the intersection is the cube. The indices [111] stand for the octahedron. To get to this, you would cut one corner each of the drawing cube in such a way that the plane of the intersection cuts the three main axes in the same ratio 1:1:1.

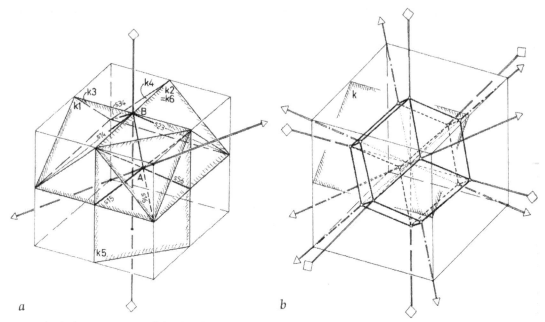

*Fig. 4.16:* The construction of the rhombic dodecahedral garnet from the octahedral-cubic coordination system and the plane k with the indices [110]: a) the construction path, b) the result.

It is characteristic of crystals that they can form composite forms; they relate the law of symmetry not only to one plane, but various planes. This shall be further explained by the example of the mineral pyrite.

### Pyrite as a Cube Octahedron and Pentagonal Dodecahedron

When studying pyrites, you will realize that they can crystallize either as cubes, as octahedral, or as irregular pentagonal dodecahedra. Even combinations of these three basic forms are common (see Fig. 4.17). Especially in the cubes and pentagonal dodecahedra, a striation of the surfaces is a striking observation and can be enhanced by etching.

This time we construct a new body from a cubic-tetrahedral symmetry, defined by four threefold and three twofold rotation axes (Fig. 4.18). 2: ∞ indices [210]. We rotate kl around the top front right facing threefolded rotation axis and get k2 and k3. The three

intersection lines s12, s13 and s23 intersect again at point A, which lies on the threefold rotation axis. This point we rotate around all three twofold rotation axes as well as around the remaining three threefold rotation axes and receive A1 to A8. The next operation rotates k1 around the perpendicular twofold rotation axis. The intersection of k1 and k4 results in an intersection line, which forms the middle of the top square in the drawing cube. The intersection between s14 with the intersection line between kl and k3 results in point B.

We rotate this point on the surface of the drawing cube around the three twofold rotation axes and yield B1 to B4. Repeat this with the intersection of k1, k2 and k5, and k5 results from a rotation of k2 around the horizontal twofold rotation axis. This results in the corresponding B5 to B8. In an additional step, we produce correspondingly B9 to B12. Our new body can be drawn by introducing connecting lines [intersection lines between two cutting planes] between one A-point to the neighboring B-points. This body contains twelve identical, irregular pentagrams. A single pentagram has a long edge [BB] and four shorter edges of the same length [AB].

You can also develop this body by treating a plasticine cube in such a way that you cute twelve wedges parallel to a cube edge to result in a twelvefold cutting face kl to k12. This pentagonal dodecahedron is not a platonic solid; its surfaces are uneven. If you want to develop a regular or platonic pentagonal dodecahedron, the first cutting face k would need to have indices in the ratio of the Golden Section. However, this would produce unnatural [non-integer] indices, which do not occur in the world of natural crystals. The gestalt of the pyrite is therefore the pyritic pentagonal dodecahedron, also called pyritohedron. When choosing a surface that is parallel to the drawing cube for a first cut, you receive a cube. A cut, which evenly cuts off a corner from the pyrite cube results in an octahedron.

When observing the oft-present characteristic striation of the cube surfaces, it becomes clear that the pyrite cube does not possess any fourfold rotation axes but only one twofold rotation axis that is oriented perpendicularly to these surfaces. This is because the pyrite cube must superimpose congruently with this striation (see Fig. 4.17a).

The pyrites drawn by students shown in Figure 4.17 show nicely that pyrite combines

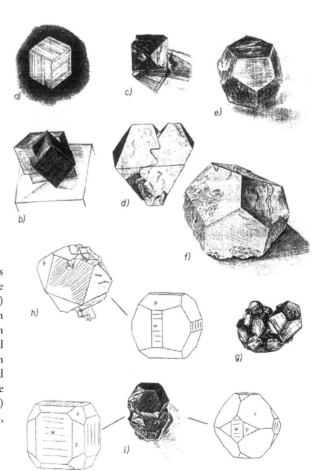

*Fig. 4.17:* Manifestations of pyrites, as drawn by students. a) Cube with visible striation. b) Intertwined cubes. c-d) Two intertwined octahedra. e) Pyrite in pentagonal dodecahedral crystallization with striation. f) Pyrite in pentagonal dodecahedral crystallization with etch figures. g) Aggregation of pentagonal dodecahedra. h) Combination of cube [w] and pentagonal dodecahedra [p]. i) Combination of cube [w], octahedral [o], and pentagonal dodecahedra [p].

predominantly octahedra and cubes, sometimes even all three forms, and thereby create highly complex spatial structures (Fig. 4.17i). It is therefore the characteristic of pyrite to play with three "first" planes. It is now important to realize that not all three planes come close enough to allow them to connect to the substance and by this become visible for the eye. If only cubes are formed, then the planes that would cause octahedra and pentagonal dodecahedra stay outside of the substance and are present only as an idea. In talking about this, students might be inspired to think that crystals could be formed from outside, through the interactions of forces in the cosmos. The conditions on earth carry the crystallizing substances to the active forces of the cosmos. As we have already seen, it is characteristic for every single substance to engage in a specific relationship with the geometrically structured active forces of the cosmos.

## Tracht, Habit and Twinning at the Example of Quartz

After a prolonged period of time dedicated to the geometrical work, it is advantageous to return to the initial consideration of the various rock crystals. Let us start with the double-enders shown in Figures 4.1 and 4.20 (Fig. 4.1a-c, Fig. 4.20a-c). The tip of the pyramid is formed by the r-face as well as one z-face rotated 60°. Both faces are treated according to the threefold rotation axis. Measurements show that the angles between corresponding areas are regularly identical. Obviously, a surface can be closer to the center of the crystal or parallel shifted further away. It can even run outside of the crystal, as we explained earlier. Mineralogists call this a distortion of the ideal crystal gestalt [shape]. This is a first aspect of the habit of a crystal, caused by the proportionality between the surface locations. It became obvious during the geometric works so far that the location of a surface in relation to the symmetry structure is always defined by ratios and never by mass. The consequence of this is the law of the constancy of interfacial angles, which also describes just the ratio of the surfaces to each other and not their position in a coordinate

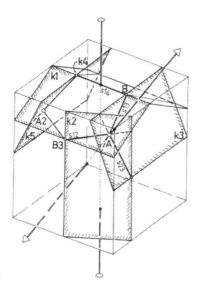

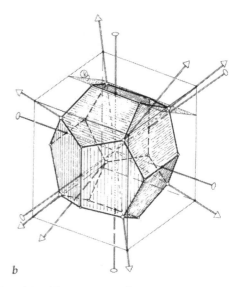

*Fig. 4.18:* The creation of a pyritic pentagonal dodecahedron from the cubic-tetrahedral ensemble of rotation axes and the cutting face k with the indices [210]. a) Construction of the corner points A and B. b) Result with the typical surface striation.

system. The distortion can be observed very well in phantom crystals (Fig. 4.1d-e).

Phantom crystals are formed if, when periodic attachments are documented, for example, the crystal receives an input of foreign matter such as, for example, chlorite resulting in a coating, grows further, attaches again foreign material for a short time, and so forth. In a section, this might look like Figure 4.19. There is obviously a different degree of attachment happening to the various surfaces. This might happen if the crystallizing solution flows around the crystal, which might cause phenomena of pressure shadows. As a consequence of the differentially strong attachment, it may happen that a surface is, so to speak, located outside of the crystal.

As we saw in the section about pyrite, crystals often form by a combination of surfaces. Caused by the various combinations, a variety of crystal "tracht" develops. The figures a, f, h, l and c in Fig. 4.1 show some "tracht" variation of rock crystals. As a common pattern, a combination of the m-, r- and z-faces appears. If M-faces appear, rock crystals become conic (Fig. 4.1h, Fig. 4.20h).

If 'x-, x'-, 'S- and S'-faces are added, you can distinguish between left- and right-hand crystals and by this discover the frequent deposits of twins (Fig. 4.1k-o, Fig. 4.20k-o). Especially the S-face is a good diagnostic helper. At the crystalline column, placed perpendicularly in front of the observer, you find the S-face either to the top right at the m-face (Fig. 4.1l, Fig. 4.20l), with a characteristic striation that runs from bottom left to top right [right-hand crystal], or other specimen show a S-face top-left with a striation running from bottom-right to top-left [left-hand crystal]. It is a striking observation that such faces predominantly occur on every next but one corner, showing

their threefoldness (Fig. 4.20k, h, o, m). In a double-ender, this surface, occurring most often as a triangle or trapezoid, can stand up to six times on six corners. It is a right-hand crystal if the trapezohedral face is located to the top right (Fig. 4.1m, Fig. 4.20m). Consequently, a left-hand crystal shows this face top left (Fig. 4.1k, Fig. 4.20k). In some of the illustrations (Fig. 4.1o and Fig. 4.20o),

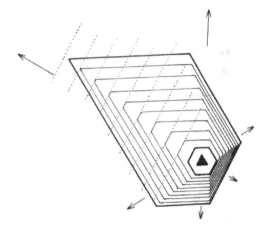

*Fig. 4.19:* Cross-section through a phantom rock crystal. Section perpendicular to the threefold rotation axis. Caused by an uneven parallel attachment, one of the six prismatic surfaces can get to be off the real crystal.

these trapezohedra are as well in left- as right-hand orientation. These are twinnings of a left- with a right-hand crystal [a so-called Brazil twin]. Other crystals (Fig. 4.1n) arrange more than three x'- or 'x-trapezohedra around a six-pyramid. They pretend to have a sixfold rotation axis. They are, however, twinnings of either two left- or two right-hand crystals [so-called Dauphiné twins]. When drawing certain specimens (Fig. 4.1p, q), students are often surprised to realize that crystals could have bent and twisted surfaces. This is called a gwindel. A right-hand crystal with a visible right-hand trapezohedron face shows a left-handed rotation (Fig. 4.1q, Fig. 4.20q). In left-hand crystals, a right-handed

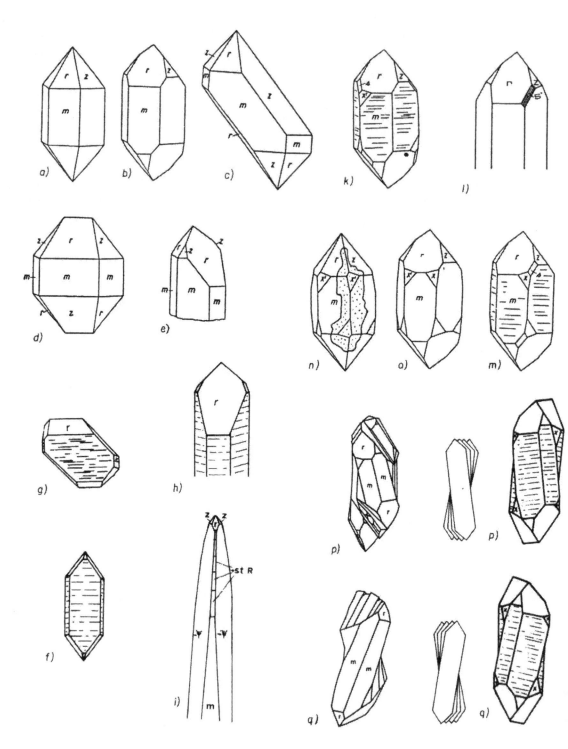

*Fig. 4.20:* Schematic representation of the variation in quartz crystals. Due to the same labeling, they can be compared to the items in Fig. 4.1. a-h) Distortions. i) Formation of a conic column due to the addition of the steep rhombohedral surfaces Z'. k) Left crystal with x-surface. I) Right crystal with S-surface. m) Right crystal with x-surface. n) Left-left Dauphiné twin. o) Left-right Brazil twin. p) Dextral left-hand crystal. q) Sinistrorse right-hand crystal. (from Rykart 1995)

rotation occurs. It is possible to say that the rare gwindel shows the character of the quartz crystal most clearly, namely the option to form either left- or right-hand forms, as this follows from the symmetry relations in its crystal. Classes of crystals without symmetry center but rotation axes are called enantiomorph, which means that they can form left- as well as right-hand forms.

This phenomenon is important when we study the interactions between life processes and enantiomorphous substances. For example, the human organism involves the laevorotatory lactate crystals in a completely different way as the dextrorotatory form. This happens despite the fact that there is no difference in the chemical composition!

With its playing of left- and right-hand forms, quartz reveals sensitivity in relation to life processes—a characteristic that is not appreciated enough by now. Further relations between tracht, habit and twinning are presented in an excellent way in the *Quarz-Monographie*. (Rykart 1995)

## The Significance of the Crystallography Block in the Context of the High School Curriculum

After providing ideas about the content and the way to conduct this block, we are now trying to recognize its significance. It is the task of the 10th grader to get away from an "if/then causality" and practice more flexible thinking and imagination. These abilities cannot yet stay in this flexibility for long periods of time. The process needs frequent periods of stopping and watching. This is exactly what we practice when we develop the seven crystal systems and relate the geometric thinking to the form phenomena in the real world of crystals and minerals.

While students look at the schematic of crystals, they experience the overall context

of one of nature's realms as complete as possible. It is especially the crystals that stand between the lifeless and the living world. Steiner indicates this clearly in the chapter "The Three Worlds" in *Theosophy*:

> In between the shapeless character of mineral matter as we meet it in gases, liquids, etc. and the living shape of the plant world, stand the forms of the crystals. In the crystals we have to seek the transition from the shapeless mineral world to the plant kingdom that has the capacity for forming living shapes.
>
>     In this externally sensory formative process in both kingdoms, mineral and plant, we must see the sensory condensation of the purely spiritual process that takes place when the spiritual germs of the three higher regions of the spirit-land form themselves into the spirit shapes of the lower regions. The transition from the formless spiritual germ to the formed structure corresponds in the spiritual world to the process of crystallization. This transition is the spiritual archetype of the process of crystallization. If this transition condenses so that the senses can perceive it in its outcome, then it exhibits itself in the world of senses as the process of mineral crystallization.
> –Steiner 1904, V./149f

The 10th grade stands actually between grasping the causality of the lifeless world and working towards an understanding of the world of the living.

When looking at the method of doing geometry in this block, you will also see a transition. On the one side you still see the Euclidian geometry with the repeated practicing of stereometric imagination. The symmetry laws, the law of the constancy of interfacial angles, the transition from measuring to ratios on the other side, points clearly to projective geometry, which will be part of the 11th grade curriculum.

The consideration of the relations between the three platonic solids—tetrahedron, cube and octahedron, entirely based on geometry—is also an appropriate and useful continuation of the earth science curriculum in grade 9, when students are pointed to the geometric spatial structure of dilatation and compression phenomena around the globe. The configuration of the double tetrahedron structure of the world (Schmutz 1986) was foreseen or consciously experienced. By adding the presented 10th grade block to the geology block, a new element of the conformity to the world's natural law appears.

This is the connection between the global structures of the earth and microscopic microstructures. The geometric motive of form creation of the geological earth is the same formal principle of the microscopic world of crystals. This is another step towards the tetrahedral principle. It should be mentioned that the methodic principle used in this block corresponds to Goethe's *anschauende Urteilskraft* [Power to Judge in Beholding]—this is the effectiveness of a greater idea, which manifests itself into the world of substance. (Schmutz 2000a)

It is still left to science to study this form principle also in the cosmos of the stars. This indicates the possibility to continue the consideration of the ratio between the Three and the Four also into the astronomy block in grade 11. This ratio should also be grasped in a new way in human science. Only this would give the crucial depth to the idea of the tetrahedral structure.

Great Sand Dunes, Colorado. Photo by DSM

# Chapter 5

# Earth Science and Surveying—Tenth Grade

## Introduction

This is not the place to speak about the importance and the workflow of the surveying block as an important part of the 10th grade curriculum. There are excellent instruction materials available, published by the Pedagogic Research Institute Kassel (Ohlendorf 1994, Ulex 1989). As a standard work in surveying, I recommend the two-volume book by Volquardts and Matthews (1975) and Kahmen and Faig (1988). The purpose of the following is to show in a few examples how important it is to work on *real projects* during the practical part of the block. Students engage much more seriously and experience real life skills during school when the surveying task carried out by the class is really useful.

## Planning of a Forest Road

For many years the Rudolf Steiner School in Wetzikon, Switzerland, was given assignments from several municipalities to survey parts of forest roads. Within the framework of larger development projects, the school took on the task to survey an entrance ramp approximately 1 km in length, entirely in a steep inclining area in the subalpine to alpine region. The road was planned to serve for the maintenance of the forest.

The preparatory work at the school centered around the practical learning about and practicing with the tools: the surveyor's level, theodolite and plane table. The following is an accounting of the seven days of work in the field and the subsequent analysis and drawing phase back at school.

### Process Steps in the Field

The overall concept was outlined in the first on-site inspection of the forest. The forester showed us the beginning and end of the road and the approximate locations for open places or turning places. He also explained which purpose the road would serve in the future.

As the approximate length and difference in elevation were known, the slope of the road was calculated on the map (with a scale of 1:10, 000 m or 1:25,000 m), and we had to plan for changes in the slope. The next step in the field was to establish the *baseline*. Stakes were fixed within the area, each 15 to 25 m apart, with a slope level [clinometers] fixed to a rod to ensure the compliance between each pair of stakes to the calculated slopes. 1000 m of road was laid out in a zigzag line of about 50 stakes. The next task was to survey this open traverse, in just one day if possible, using several theodolites. In the evening, the traverse was drawn in a scale of 1:500.

Next was the elongation of the baseline (Fig. 5.1). The drawn baseline corresponded approximately with the planned run of the road, but it had too many curves. So a simplified zigzag line was designed, which did not deviate much from the baseline. The so-called *traverse by tangents* (Fig. 5.1) contained as many tangents as necessary curves.

On the next day the traverse by tangents had to be marked in the field. The planned tangent points could not be marked where they should be in the field because there needs be a line-of-sight link between each point for working the theodolite! Unfortunately, there were too many trees and bends in our way. If the tangent traverse was already staked out, it had to be accurately re-measured because these tangent points would become the jacket of the new road.

In the evening, the corrected tangent traverse was recorded and the curves, including a curve widening, was projected according to the initial instructions (Fig. 5.1). This way we got approx *100 axial points*, which marked the middle of the planned *axis*.

After the axis was properly marked the next day, using stakes which were placed with the help of surveyor's tape and optical squares, the leveling groups, who had already measured the tangent points, could determine the height of each of the hundred stakes. It was important to work carefully and efficiently but at the same time avoid any transfer of accidental errors.

A second group determined the exact horizontal distance between the stakes using surveyor's tape and staff, building the base for the preparation of the longitudinal profile (Fig. 5.3).

That evening, the exact course of the road axis was drawn using the angular and distance data. In various places with more complex topography and planning conditions, other groups created a topographic map including the placed stakes in a scale of 1:200 using *plane tables* in the field (Fig. 5.2). This was done in the area where the planned forest road connected to the existing street, in an area of a difficult turning space, and at the end of the road where a cul-de-sac was required. At each axial point, a *cross-section* 10 m left and right of the axis had to be developed using four-meter staffs, range poles, water level and double meter stick (Fig. 5.5).

One last local inspection along the staked axis served to note any specific terrain conditions such as rocks, landslides and moist patches or gullies that would require drainage before construction.

### Analysis and Quantity Surveying

Following the surveying camp, the main task of phase three of the project was to do the calculations. From the length and height data of the road axis, the *longitudinal profile* was drawn and stretched. On a trial basis, the *line of elongation* was positioned in such a way that the axial points were not too far away from the line.

Changes in the slope of the calculated line were connected using a parallel curve (Fig. 5.3). It could be read from the drawing how many centimeters the road would run above the real terrain [dam] and below the terrain [cutting]. The planned road cross-sections were drawn into the cross profiles together with a ballast roadbed and excavated or deposited banks (Fig. 5.4). The *deposition and excavation volume* can be calculated from the area cross-sections of two profiles and the distance between them. The task was now to place the elongation line and the corresponding volume calculations on a trial-and-error basis in such a way that deposition and excavation volumes would even each other out. Additionally, it was important to make sure that the transport volume was transported on the shortest route possible.

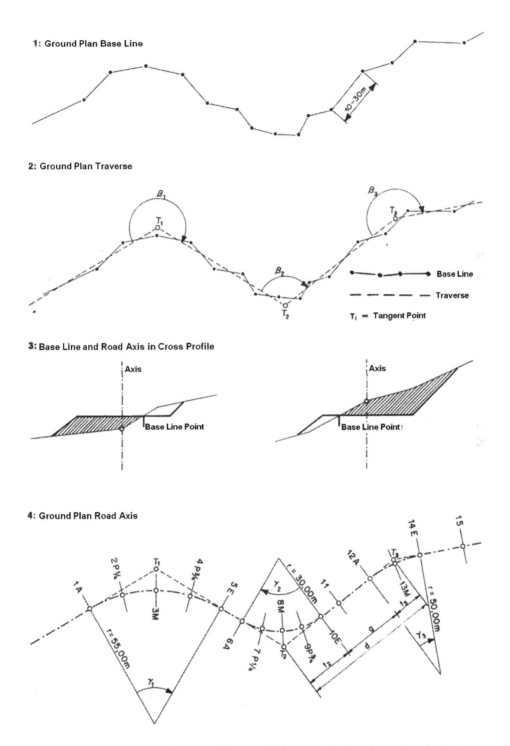

*Fig. 5.1:* Graphic planning steps in the project planning of a forest road. Plot 1 shows the staked and measured baseline. Plot 2 shows the elongation of the baseline to simplify the curve shape of the finished road. The cross-sections in Plot 3 demonstrate what will happen when the road deviates from the baseline: If the road lies downhill from the baseline, it will need more embanking; an axis that is shifted uphill would result in a larger cutting. Plot 4 shows the axial points with the graphically determined path elements [A = springing, M = middle of the arc, E = end of the arc, r = curve radius]. (From Swiss study group for forest road construction [Schweizerische Arbeitsgemeinschaft für forstlichen Straßenbau], Merkblätter 102–200, 1977)

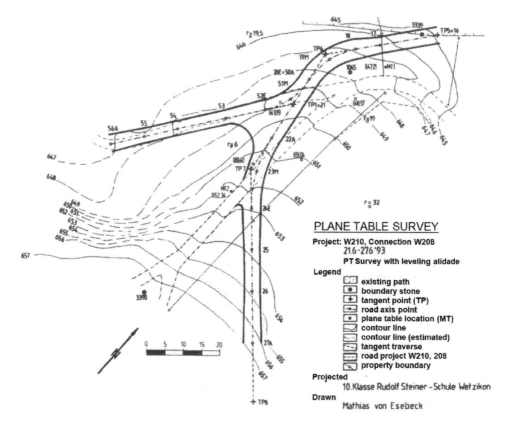

PLANE TABLE SURVEY

Project: W210, Connection W208
21.6-276'93
PT Survey with leveling alidade

Legend
existing path
boundary stone
tangent point (TP)
road axis point
plane table location (MT)
contour line
contour line (estimated)
tangent traverse
road project W210, 208
property boundary

Projected
10. Klasse Rudolf Steiner - Schule Wetzikon

Drawn
Mathias von Esebeck

*Fig. 5.2:* Example of a plane table survey of a section of the area with a junction of two projected road fragments. In this section, the cross profiles could be read from the map.

This was the purpose of drawing a mass profile (Fig. 5.5). This lavish work made sure that the new road could be constructed with the least impact on the landscape and the most cost effectiveness.

After the ideal line for the elevation was determined, all maps and plans had to be drawn and labeled according to the regulations. The most exact and beautiful maps were handed over to the developer (see Fig. 5.6).

All tasks described were done as a group effort. A special event was always the terrain inspection of the construction site, during which the class saw how the foreman directed the machinery according to the maps they had developed.

### Epistemological Note

Using notes from *The Science of Knowing: Outline of an Epistemology Implicit in the Goethean Worldview* (Steiner 1886), chapter 1 explains how the process of knowledge develops from the matching of the given perceivable to the added conceptual. In this, the universal-ideal conceptual is specialized and paralyzed to an imagination.

In the alternation between fieldwork in the day and office work in the evenings, the student observes how an idea is related step by step to the perceivable of the given terrain. The idea, in this case the road project, has to be modified and specialized as much as it becomes compatible with the given of the topography. In this way the general idea becomes something that can develop into the

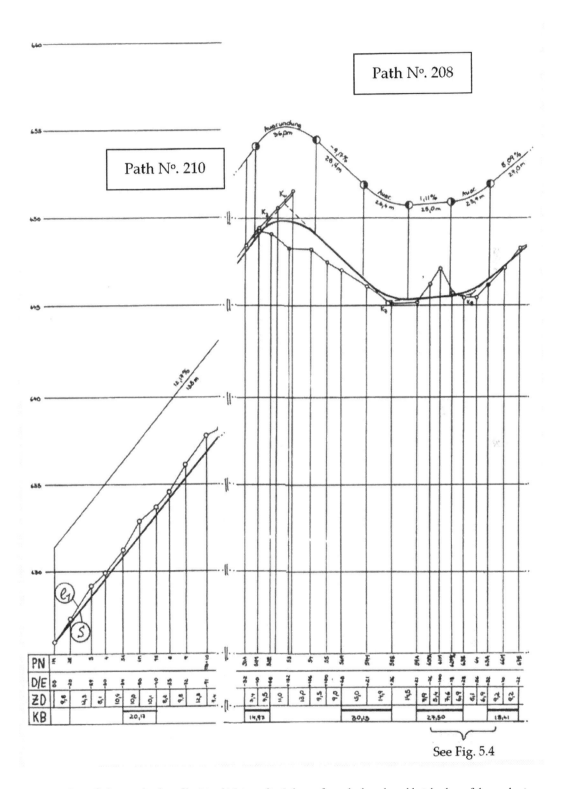

*Fig. 5.3:* Part of a longitudinal profile. Line l1 [zigzag line] drawn from the length and height data of the road axis. The elongation [line S] was placed in such a way that deposition and excavation volumes were approximately the same (Fig. 5.5). PN = profile number, D/E = dam/cutting in cm, ZD = distance in between in m, KB = curve belt.

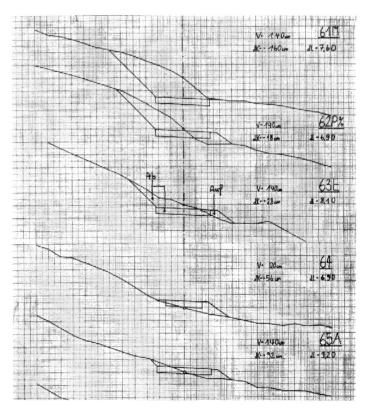

*Fig. 5.4:* Detail of the cross profiles with corresponding quantity surveying. The excavation [Ab] and deposition areas [Auf] were determined by counting the number of squares on the plotting paper. V= road widening in curves; dk = difference between planned road surface and axial stake in terrain (comp. Fig. 5.3); dl = distance between two cross profiles (see Fig. 5.3)

**Earth Mass Calculations**

Projekt: Wila 1993 H 210/208     Seite: 2

| Profil Nr | Area of the Cross Profiles | | Area of each two cross profiles | | Distance in-between | Cubic content between each two cross profiles | | Use of Earth Masses | | | |
|---|---|---|---|---|---|---|---|---|---|---|---|
| | Excavation | Deposition | Excavation | Deposition | | Excavation | Deposition | On site | With transport | | |
| | | | | | | | | | Excavation | Deposition | |
| | m² | m² | m² | m² | m | m³ | m³ | m³ | m³ | m³ | |
| 1 | 2 | 3 | 4 | 5 | 6 | 7 | 8 (Zuschlag 30%) | 9 | 10 | 11 | 12 |
| 60PY4 | 12.1 | 0.0 | 14.9 | 0.9 | 7.90 | 58.9 | 4.6 | 4.6 | 54.3 | | 12 54.4 |
| 61M | 13.0 | 0.0 | 25.6 | 0.0 | 5.40 | 69.1 | 0.0 | 0.0 | 69.1 | | 13 25.5 |
| 62P3/4 | 6.2 | 1.3 | 19.2 | 1.3 | 7.60 | 73.0 | 6.4 | 6.4 | 66.6 | | 13 92.1 |
| 63E | 4.9 | 0.8 | 10.9 | 2.1 | 6.90 | 33.6 | 9.4 | 9.4 | 28.2 | | 14 20.3 |
| 64 | 0.4 | 1.6 | 5.1 | 2.4 | 8.10 | 20.7 | 12.6 | 12.6 | 8.1 | | 14 28.4 |
| 65A | 1.1 | 0.9 | 1.5 | 2.5 | 6.90 | 5.2 | 11.2 | 5.2 | | 6.0 | 14 22.4 |
| 66M | 1.3 | 0.2 | 2.4 | 1.1 | 9.20 | 11.0 | 6.6 | 6.6 | 4.4 | | 14 26.8 |
| 67E | 5.7 | 0.1 | 7.0 | 0.3 | 9.20 | 32.2 | 1.8 | 1.8 | 30.4 | | 14 57.2 |
| 68 .... | 1.3 | 0.2 | 4.0 | 0.2 | 9.30 | 19.6 | 1.3 | 1.3 | 18.3 | | 14 84.3 |
| 69A 36M | 2.9 | 0.0 | 5.4 | 0.0 | 9.40 | 25.4 | 0.0 | 0.0 | 25.4 | | 15 02.6 |
| **Control of areas** | 81.8 | 45.3 | 160.7 | 87.9 | | 239.70 | 869.0 | 382.9 | 2014.1 | 486.1 | |
| | ·2 | ·2 | + 0.2 / + 2.7 | + 2.7 / 0.0 | | | | | +382.9 | +382.9 | +486.1 |
| | 163.6 | 90.6 | 163.6 | 90.6 | | **Control of masses** | | | 2397.0 | 869.0 | 2014.1· |

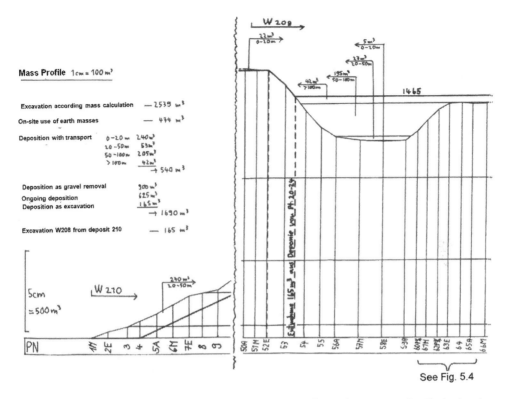

*Fig. 5.5:* Part of a mass profile. The transport of earth masses along the road axis are graphically displayed in a longitudinal profile. Part of the excavation volume is shown as depositing material in the form of directly usable gravel. PN = profile number.

world of works deeds. The various planning and measuring tasks serve to adapt the idea to the conditions of the terrain and the specific functions that the road will have to serve. The student therefore experiences the same process which he will be able to observe as an adult mentally during the act of knowing. As in the crystallography block (comp. chapter 4), the same relation appears between the ideal and the world of deeds. The ideal gestalt context of the tetrahedron is modified and specialized step by step—thus founding the basics to describe real crystals into a system.

## Documentation of Landscape Design for the Preservation of Historic Buildings and Monuments

For several years, the Rudolf Steiner-School Wetzikon was given the opportunity to survey historic Renaissance and Baroque gardens in the Bergell region within the framework of research projects for the cantonal preservation of historic buildings. The established maps in a scale of 1:100 and 1:200 served as the documentation and planning base of urgently needed renovations. Students got to know a small piece of humanly-influenced mountainous region up into the small details.

### *Surveying of the Salis-Palazzi Gardens in Soglio*

The Baroque gardens in Soglio were surveyed using plane tables (Fig. 5.7). This task demanded quite a bit as the terrain of the mountain village in the Grisons was steep and difficult to overlook. Individual plane table maps were combined using survey traverses. By doing this work, students got to understand and appreciate the very subtle landscape design. On the one side, the spatial orientation of the walls and their scaled heights are in accord with the seasonal extremely changing path of the sun. On the other side, the architecture of the walls increases the experience of the landscape panorama of these mountains as seen from inside. Students were able to experience by their own observations how nature can be lifted to a higher level with the use of sensible cultural achievements.

### Survey of Medieval Grotti

A highly rewarding assignment was the documentation of the medieval grotti in the Italian part of the Bergell region (Fig. 5.8). In this area, full of nooks and crannies, with vine cellars framed into a rock slide area, the challenge was to conduct exact measurements using a theodolite, leveling instrument and plane table, and combine them with small-scale measurements, taken where only an optical square, staff and double meter could be used.

It therefore had to be decided, on a case-to-case basis, which measurement technique to apply. The composition of the partial results required a high degree of teamwork and overseeing of the successful work of other groups of students. This promoted a social work community of the students, which amazed the local inhabitants each day more and more. These people understood how to reach a high level of living standard by using the most simple rock architecture.

The question of why vine cellars were built especially in the area of big boulder rockslides could be answered by observation. In each grotto, the back wall in the basement consisted of large rockslide boulders. Year-

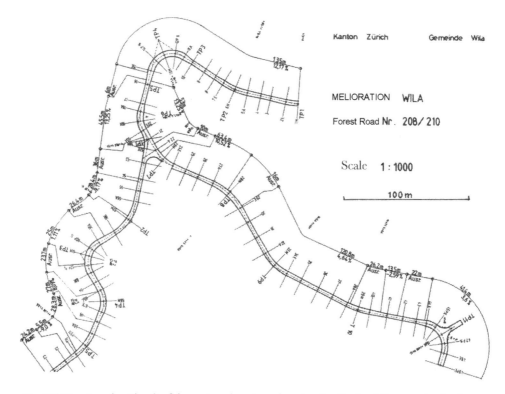

*Fig. 5.6:* Situation plan [detail] of the projected road; student drawing for the delivery to the developer

98

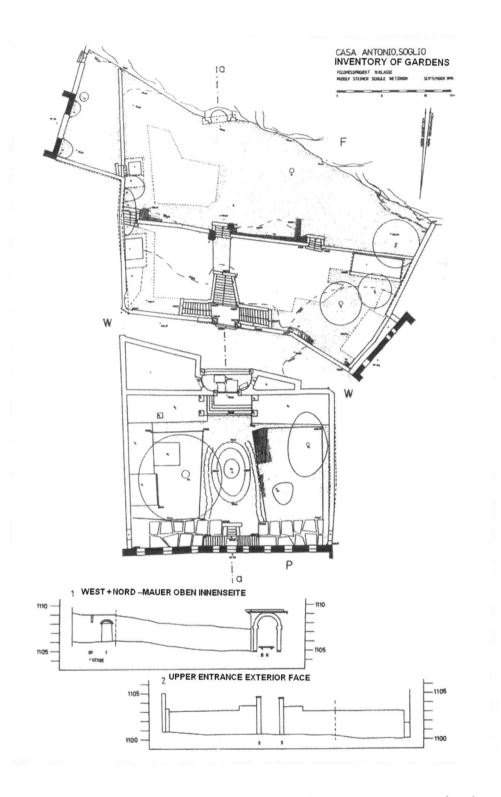

*Fig. 5.7:* Student drawing of a surveyed Baroque garden in Soglio. The gestalt composition strives from the ratio between the axis a-a, which is perpendicularly oriented to the Palazzo Casa Antonio [P], the rocky area in the north [F] and the public cut through path [W], which the farmers in the village had to fight for to get. The wall elevations are shown only in parts.

round, a breath of air with a temperature between 8° and 9°C flows through air gaps between the stones—creating a premium natural cold storage.

## Pedagogic Considerations

It proved beneficial to conduct the surveying as a focused three-week project, following the math block of trigonometry. The first week of the project was on the school grounds and served to acquaint the students with the equipment. The second week was focused on work in the field, and evenings were spent for necessary calculations and drawing. Back at school we finished the project drawings and calculations to hand over to the ordering institution, and we wrote up a final report. This was an exemplary way to learn to prepare the results of our work in a way that could be used by other people to conduct further tasks. Over time errors and gaps had to be corrected and filled in. The self-esteem of the class was high when, astonished, the public authorities expressed praise. Social tensions were lessened. The following insight developed: Human societies develop cohesion by taking responsibility for something which can then be put to somebody else's disposal for implementing.

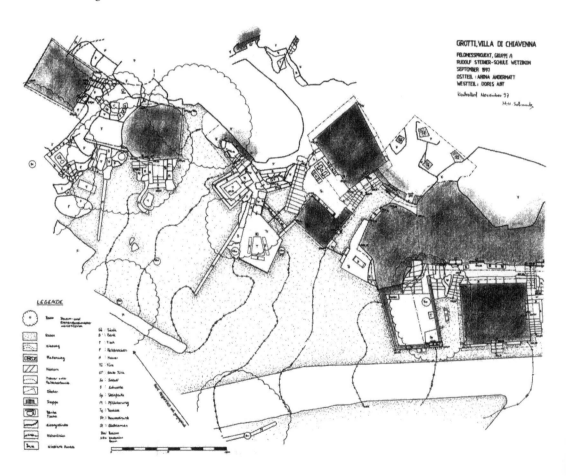

*Fig. 5.8:* Student drawing of the surveyed grotti in Villa di Chiavenna, Italy (selection): Plane table survey including contour lines in meter intervals; construction volume indicated with gray shading; rock boulders are surrounded by thick lines.

# Chapter 6

# Aspects of Life Skills, Technology and Economic Geography for Grades 9 and 10

## General Conditions

The earth science blocks in grades 9 and 10 focus on geology, mineralogy and physical geography. The question arises also where to place economic geography and cultural geography. If selecting appropriate projects, the surveying block (chapter 5) could be part of cultural geography. The history block in grade 10 is also strongly influenced by cultural geography. In his lectures on public pedagogy, Steiner demanded very clearly *instructions in life skills and technology:*

An understanding of life must inspire all teaching. We need to rationally and efficiently teach students from fifteen to twenty years of age, everything having to do with agriculture, commerce, industry and trade. No one should pass through this period without acquiring at least an idea of what occurs in agriculture, in trade, in industry and in commerce. These things need to become individual disciplines and are much more necessary than much of the rubbish that now fills education during these years. At this time in life, we need to teach all those things I would generally call "world affairs." Among these are

history and geography and everything connected with an understanding of nature, but always as they relate to the human being, so that children learn about human beings in a universal context. (Steiner 1919d, first lecture, 5/11/1919, "Education as a Force for Social Change," *Foundations of Waldorf Education, vol. 4*, SteinerBooks, 1997)

Aspects of economic geography can be included naturally within the lessons in technology or life skills. These lessons can either be distributed over the entire school year and enhanced with *periodic tours of factories and plants*, or a second block or so-called crafts blocks can be allocated for this purpose. In my school we had crafts blocks that lasted four to seven weeks each and were taught four to six hours per week.

For this block, no main lesson book had to be worked on outside of class. Instead, the working technique is learned: to take clear and relatively detailed notes during the lessons, which are collected in a workshop manual. The students also learn how to deal with statistical data and press releases. The whole path of instruction is more practice-oriented.

## Visiting Factories and Plants in Grade 9

Experience shows that it is most beneficial to visit the service industry in grade 9. One representative example is described below in detail.

The visit to the garbage incineration plant Zürcher Oberland was a rewarding trip, especially due to the director, who was one of the pioneers of the concept of waste separation in private households. The students met a human being who did not only fulfill his tasks, but actually almost worked against his own enterprise by fighting for the least amount of garbage possible to be burned in his plant. This way the region Zürcher Oberland was a good example: Regional composting, strict waste separation by the consumer and the passing-on of the separated materials to recycling companies has been working well for many years.

The incinerator itself was well worth even a longer visit. Students could follow the path of a garbage bag from the intermediate storage in a bunker to the charging into the furnace. The process of combustion could be observed through special peepholes. The incombustible slag could be inspected on a belt conveyor. The produced heat was first converted into steam and powered an electric generator; the water was cooled down to 90° C and served for the distant heating of the surrounding industrial plants.

Such tours of factories and plants serve several purposes: They allow the students opportunities to examine today's civilization processes and understand the functioning of such plants. Students should experience that facilities work well when the inventive and realizing genius of human beings is activated. The initiating and running of such plants is achieved through an appropriate thinking. Thirdly, the young people awaken in respect to their own actions as ordinary consumers. It happened very often after visits to the garbage incineration plant that garbage separation units were established in the classroom. Parents reported that their sons and daughters took care of the garbage separation at home, at least for awhile. It is typical for 9th graders that they want to practice immediately what they recognize to be the right thing to do. In this case, this is easily done.

Additional tours depend on local opportunities: a mail distribution center, a bread factory of a large food chain, a paper mill, the city gas plant. Everywhere it can be experienced how human beings work in tedious jobs to allow other people to enjoy relatively comfortable day-to-day lives.

## Textile Technology in Grade 10

After students have gone through needlework classes during eight years of lower and middle school and a tailoring block in grade 9, a transformation into a technology class is indicated for grade 10. Manual skills were practiced in many ways. Now it is important to move from getting-to-know things and skills to more discovery-based learning experiences, hence the term "technology." This transition is begun with the *spinning class*. (Karutz 1991a; 1991b) Manual skills are further trained and an increasing mental understanding of the work processes is added. Students appreciate the congenial inventions from the hand spindle to the hand spinning wheel and the treadle wheel in hands-on doing. (Hentschel 1975)

In review of the spinning block, the *textile technology class* might begin with socio-economic aspects as we will specifically investigate below. Good sources for an understanding of the developments in textile technology are the books *Spinning and Weaving*, Bohnsack, (1981), as well as Burnham (1981), English (1969), and Hecht (2001).

Steiner referred to the importance of rivers to the study of the development of human culture and trade even for middle school geography classes. This key shall be used again with the example of the industrialization in the textile trade. As an industrial region, the Zürcher Oberland [Zurich highlands] was always characterized by the textile industry. After the first pioneer plants were developed in England, the first textile factories on the continent were established in the Zürcher Oberland. Most of this is already known to the students from their local history, geography, and natural history classes in the lower school. Many old buildings from this period remain part of the landscape but today are often used for other purposes.

Industrialization took place in this region due mainly to two conditions: a previous presence of the textile craft and the availability of rivers. In the hilly Zürcher Oberland, farmers always needed a second source of income. On each homestead there was a weaving loom, and several spinning wheels stood in damp basements. In domestic work, wool, flax, hemp, and cotton were spun and processed into fabric. Over centuries, elaborate textile skills and a well-functioning trade organization were established. After the invention of the first machines in England, people quickly realized how favorable the existing plenitude of small rivers with appropriate downward gradient was for the use of waterpower-driven machines. Due to this, many technically innovative strategies of energy conversion were developed via spinning and weaving mills.

It is interesting to track how the occupational situation of spinners and weavers changed due to developments in technology. (Bohnsack 1981) As the social consequences of these technological revolutions were not reduced, wage cuts and unemployment alternated between spinners and weavers. Simultaneously, impressive and oppressive are the portrayals of child labor in factories. As in the example of the industrialist Guyer-Zeller, social inventive genius and ethical behavior slowly implemented social achievements such as health insurance, social welfare, humane working hours, social housing and other benefits. The ongoing reduction of working hours paralleled the construction of more powerful and efficient machines. (Berner et al. 1962)

In class, portrayals of the open-ended technology can be continued up into present time. Key phrases appear such as: low-wage countries, globalization of the world economy, factory robots and the impoverishment of the so-called less-developed countries. But these modern themes cannot yet be processed; they are, if at all, reserved for the economy block in grade 12.

Why of all topics is it precisely the textile industry that is covered in grade 10? Based on historic investigations, we can show that the entire story of the development of industrialization happened in the textile sector. The first small manufacturing and factories occurred in textile processing and finishing. The chemical industry developed out of textile dyeworks, the machine industry started with the construction of textile machines, and modern energy conversion machines like the steam engine were first used in the textile sector. Practical experience has shown that it is highly beneficial to implement Steiner's suggestion of the principle to review the steps of the development of mankind once more. The 11th grade is the right time to study energy conversions (chapter 8), and grade 12 takes an exemplary look at the production and trade of essential foods (chapter 10).

# Chapter 7

# Astronomy—Eleventh Grade

## Preliminary Remarks

It is the intention of the present chapter to introduce the reasons an astronomy block should follow the study of man, and to provide suggestions for the design of this block. For the basic scientific principles, you should consult primarily *Rhythms of the Stars* (1986) and Davidson (2004), and consider observing the starry sky itself, with the help of periodically published star calendars such as the one from the mathematical-astronomical section of the Goetheanum [Freie Hochschule für Geisteswissenschaft Dornach], the Abrams-Planetarium and/or the Griffith-Observatory [both in the references]. An alternative approach to this 11th grade block was published in a special issue on astronomy of the German educational journal *Erziehungskunst*. (Haag 1998) In addition, there are two German publications on astronomy through the grades by the astronomer and high school teacher Schmidt. (Schmidt 1998; 1987)

## Epistemological Principles of the Concept of Intentionality [Intentional Relation]

As will be shown in the following sections, the 11th grade practice theme is the awakening to handle intentional events. Let me first explain the key concept of *intentionality* [intentional relation].

Within the epistemological part of the introduction, an example was used to explain the four stages of the formation of reality, starting with the prototype of ideals and leading to an imaginable, solidified world of deeds (refer to chapter 1). Most difficult to understand is the second stage, representing the transition from an idea to a life process adapting to the conditions of the real world. Steiner dealt with and elaborated on Brentano's concept of intentionality in *Von Seelenrätseln*. (Steiner 1917, appendices 5 and 6, translated in part as *The Case for Anthroposophy* by Owen Barfield 1978) Witzenmann described intentionality [the intentional relation] in a series of essays called "4 x 12 and 3 x 7: The on-looking power of judgment and being aware sensually and morally." These essays follow the four causes of Aristotle [the causes of substance, form, movement and purpose] resulting from observations of mental and emotional processes: When merging the perceivable incoherent and the conceptual coherence, the creation of reality occurs in four steps. After the mental performance of grasping a universality, which is called purposeful activity corresponding to Aristotle's purpose cause, Witzenmann described the next stage as follows: "Because of this generality of the purposeful activity, it needs to be adjusted to the specific implementation conditions, into which it will be incorporated. A purposeful activity (1) must therefore be joined by a movement (2), which aligns the general to the direction of the non-general [the individual]. The *moving and shaking intentionality* [purposefulness] is *intentional in a double sense* [italics by the author]. The recognizing human being, on the one side, can intentionalize his current universalities towards certain riddles [not yet recognized contexts] on a trial and error basis. In the multitude of world phenomena on the other hand, certain universal formative powers are not incorporated into arbitrary formations but only into those that they are [intentionally] associated with. This intentional association alone does not lead to an object-forming realization. On the contrary, it needs completion through a forming and formable adaptability. It expresses itself by the mobility of universalities [general concepts]. And they can adopt numerous forms, e.g., the general concept of a fir tree can be imagined in numerous variations of specific trees. Universalities are therefore formable. But they are also form-giving precisely because of their formability [plasticity], because they provide the objects with form due to their adaptability. (3) The forming formable ability of metamorphosis can therefore be listed.

But even this does not yet lead to realization. For this it is necessary to arrest (4) the adaptability of universalities through a *materially formable and forming transformation* in a certain state of formation. This happens if a general concept is arrested through a field of perceivable non-generalities and by this receives an individual form. This process of the individualization of a universality [or several universalities which are all specifically representative for everything] by non-universalities, which is inherently a formative-empowered system into formlessness, usually takes place in the subconscious of the realizing human being. The proof for these processes is our imagination. These are individualized concepts that can be remembered independently of the process of recognition as its subsequent results and evidence. We can therefore outline a process of recognition in four continuously merging steps. The sequence of steps corresponds with Aristotle's four causes: 1. Purposeful activity, 2. Moving intentionality, 3. Formative formable metamorphosis, and 4. Dissimilar malleable inherence.

One possibility for detecting the soul quality of the intentional event as an interaction is the observation of the cooperation between two human beings. The formative momentum of one human being affects the other human being. The latter treats this momentum as he can and transforms it. The event causes a reaction from the first person. In addition, he/she also receives the formative momentum of the second human being. The interrelation between these formation momentums cannot be explained with the following pattern: A certain effect follows mandatory on a certain cause. On the contrary, the "cause" becomes "effect" and vice versa. The intentional event is therefore taking place in the spiritual-soulful; it is an *interrelationship between creators*.

In the act of wanting, the human being turns towards the world. In the act of feeling, man experiences the relation of the world to himself. In the soul life of man, the formation of an imagination or a concept is the purpose of an action. Such imaginations can originate from one human being and can be perceived and adopted by another human being. In chapter 9 of *The Philosophy of Freedom*, Steiner proposed that this imagination only rises to become the motivation for an action if it meets a suitable driving force, or, in other words, only if there is a positive encounter with the characterological disposition of the human being:

> In any particular act of will, we must take into account the motive and the driving force. The motive is a factor with the character of a concept or a mental picture; the driving force is the will-factor belonging to the human organization and directly conditioned by it. The conceptual factor, or motive, is the momentary determining factor of the individual. A motive for the will may be a pure concept, or else a concept with a particular reference to a percept, that is, a mental picture. Both general concepts and individual ones [mental pictures] become motives of will by affecting the human individual and determining him to action in a particular direction. But one and the same concept, or one and the same mental picture, affects different individuals differently. It stimulates different men to different actions. An act of will is therefore not merely the outcome of the concept or the mental picture, but also of the unique make-up of the person. Here we may well follow the example of Eduard von Hartmann and call this individual make-up the characterological disposition. The manner in which concept and mental picture affect the characterological disposition of a man gives to his life a definite moral or ethical stamp. (Steiner, 1894, chapter 9, sect. 7, 149 f., Seventh English edition, translated from the German and with an introduction by Michael Wilson)

In the affairs of the world, this soulful, directing, determining and being determined activity happens within the interactions between the cosmos and life on earth. Steiner dedicated the third scientific lecture cycle for teachers to this topic: "The Relationship of the Diverse Branches of Natural Science to Astronomy." (Steiner 1921) He showed there that biological facts, especially embryology, cannot be understood without astronomy. In return, astronomy will be enlightened only by the investigation of the nature of earth and the human being.

## The Sequence for the High School

The overall curriculum of the four grades in high school was characterized in chapter 1. After students work through the understanding of the sensual work of deeds (grade 9) to an understanding of the processes of life (grade 10), the question is about the formative momentum of life processes. The question *Why?* is asked more clearly. This question turns the view toward the purposefulness of life processes; these are, in other words, intentional events.

This practicing of intentionality may originate from two different perspectives. One possibility is to look at the earth and study the intentions of human actions towards and with the earth. The economical technology block (chapter 6) provides a good opportunity to deal with these questions: Which courses of action make sense, which do justice to the human being, which to nature? (Compare this to block reports by Bechinger and Kübler, in Kübler 2000a.) The other view looks outside, into the cosmos. Here the question is: In which relation to each other are cosmos and earth? Astronomy is the appropriate block to address this question.

## Purposes of the Astronomy Block—Basics about Grade 11

This is not just a block on celestial mechanics and/or the mathematical grasping of the cosmos. At the core is first the concept of *rhythm*, a concept key to understanding life. This should definitely be based upon the practiced concept of causality (grade 9: If one thing collides with another, a definite third happens) and condition (grade 10: If one thing leads to another, the first thing *might* develop further). Practicing intentional events is now in the center of 11th grade studies. When looking into the world, it is first the astral world [the cosmos] that corresponds to the previously stated. In the above-mentioned *Third Scientific Lecture Cycle*, Steiner presented very strongly to the teachers that the conscious taking-up of the relations between astronomy and the various areas of the living should be the task of the present time. The importance of astronomy can be understood only if one also considers life on earth [e.g., embryology, evolution, human history]. We may not overlook the following: You can relate things or beings to each other only if you know and understand them individually! Therefore we will first have to work on the basics of astronomy or, in the most favorable case, pick up from where we left off in middle school. It is highly beneficial to allow students the following experience: In grasping cosmic-earthly relationships, I can get a feel for the fact that modern scientific ideologies [the human being as an accident, the human being as a causal necessity of x, the human being as a series of merged catastrophes] do not need to be adopted—because there are other ideologies to be developed. These other worldviews will be taken up in grade 12.

In summary, the following can be seen as a guideline of this block: We can understand and later grasp the earth as well as human activity and work on earth only if the relations between cosmos, earth and man are discussed in a concrete way. An *ethics of insight* in relation to human activity on earth requires essentially also an *understanding of the nature* of the interactions between cosmos and earth. The connection with the cosmos, which was temporarily lost in the history of mankind as well as during adolescence, should be reinstated by means of practicing observation and thinking. This will be illustrated by the examples in the next sections. This should also make clear why the astronomy block has to precede the block on the economy of energy.

### The Course of the Block

#### Sun—Earth—Stars

In the following, some specifications about the content are combined with methodological hints in order to show how such a block could be designed. First it is all about grasping the *space-time relations* in the solar system in such a way that it leads to the concept of *rhythm*. The consideration of the earth-sun path will fill the first week of the block classes. The path of the sun as observed from our subjectively stationary standpoint on earth leads to several different rhythmic interwoven motions.

Slight day-to-day modifications of the sun's path lead to seasons. The three main daily paths [winter solstice, equinox, summer solstice] are shown in Figure 7.1 in such a way that I watch myself as an observer with my horizon around me from far outside in the cosmos. What is shown from outside is what I see inside. In addition, the sky around me is characterized by the pole star. The celestial equator, the daily path of the sun during the equinoxes, is oriented perpendicularly to the

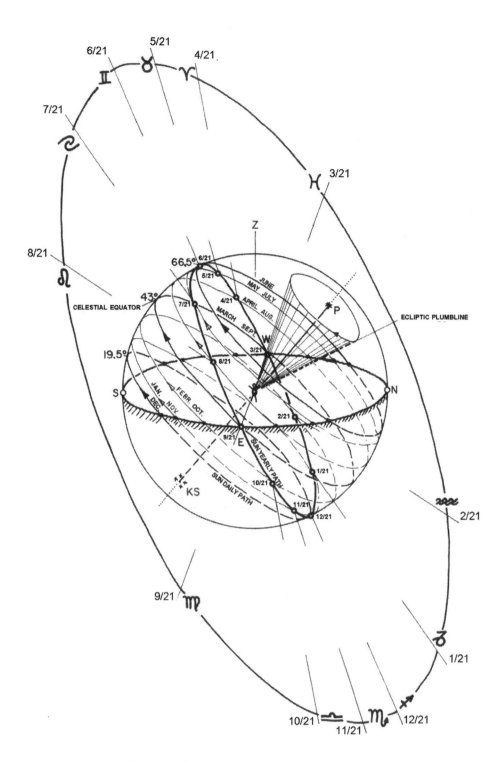

*Fig. 7.1:* Geocentric view of the daily and yearly paths of the sun with the zodiac and the ecliptic. What is shown from far outside is what the observer sees from inside. The position of the zodiac is shown for March 21 at 6 pm, and is pulled out of the celestial sphere for graphical purposes. The yearly path of the sun as the ecliptic [bold line] was developed graphically by plotting the position of the sun month by month at 6 pm astronomical time. The normal to the ecliptic plane [perpendicular to the ecliptic] rotates daily and yearly around a lateral surface of a cone with 23.5° aperture angle (comp. Fig. 7.3). P = Pole star, KS = Southern Cross, Z = Zenith

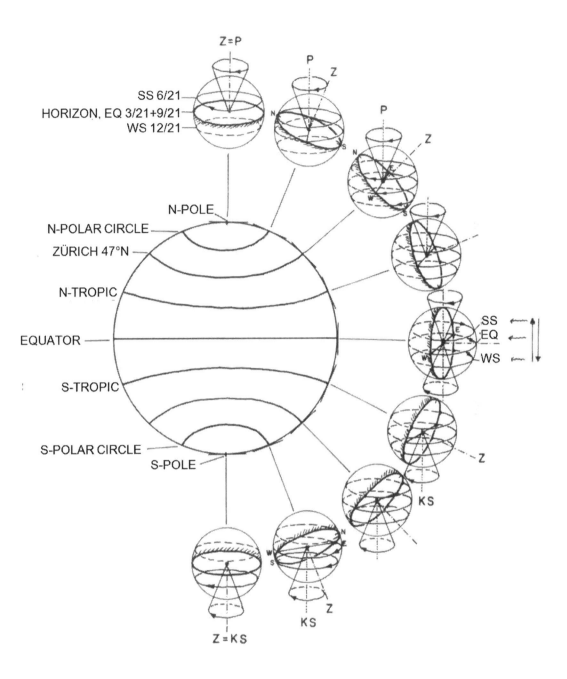

*Fig. 7.2:* Representation of the sun's path on 3/21, 6/21, 9/21 and 12/21 for various locations of the observer between the North Pole, Equator and South Pole. In terms of the yearly rising and setting of the sun, the earth is imagined as being stationary. Individual pictures are assembled in such a way that the daily path of the sun runs horizontally in all pictures.

arc between the observer and the pole star. Based on the yearly observation of the sun's path, such a representation can be acquired at various locations and latitude.

The corresponding cone marks the daily and yearly rotations of the perpendicular to the ecliptic on the cone surface [seesaw motion of the ecliptic]. The position changes of the horizon points to the spherical shape of the earth. EQ = Equinox; SS = Summer Solstice; WS = Winter Solstice; P = Pole Star; KS = Southern Cross, Z = Zenith.

In Figure 7.2, the pictures that were taken at each time are assembled in such a way that the daily paths of the sun stay more or less parallel. In return, the horizon rotates from its horizontal position at the pole to its vertical position at the Equator. When you add the fact that the variation in the latitude does not alter the representation, Figure 7.2 becomes evidence for the spherical shape of the earth.

Let us continue to discuss the valuable information in Figure 7.1. For graphing reasons, the positions of the signs of the zodiac on March 21 at 6 pm were drawn onto a ring outside of the celestial sphere.

## The Discovery of the Ecliptic as the Yearly Path of the Sun

The starting point is the fact of the various observation periods of the movement of the sun and stars. While by convention, the sun needs 24 hours for one period of rotation, a star rotates once through a *sidereal day* in approximately 23 hours and 56 minutes. Starting on March 21 at 6 pm, and referring to *astronomical time*, the position of the sun is progressively plotted from month to month. One month later at 6pm sidereal time, the sun is in the west-southwest and about 20° up in the sky because it limps behind four minutes per sidereal day. On June 21 it is 66.5° high in the sky above south. This way

we graphically yield the *sun's yearly path* (bold line in Fig. 7.1). The observer stands in the center of this circular route [orbit], which is more or less diagonally wedged in between three extreme daily sun paths. The zodiac stays in the same position because the position of the sun is always shown at the same time [pm sidereal time].

Another thought leads to the conclusion that the position of the *ecliptic* changes during a sidereal day. Let us look at the situation on March 21 at 12am: The sun is at the lowest position of its daily path in spring. The ecliptic must therefore also run through this point. Figure 7.3b shows the ecliptic as determined from the sun's month-to-month position always at 12am sidereal time. A corresponding construction of the ecliptic position at 6am and 12pm is shown in the Figures 7.3c and 7.3d. Now the result is final: The ecliptic swings through all four positions in a day. The same happens with the zodiac because it is the continuation of the ecliptic towards the outside.

But how about modifications of the ecliptic position over the seasons? We can also draw the ecliptic position over the different seasons starting with the situation shown in Figure 7.1. We need three points to define the plane of the ecliptic. The first point is always given, the location of the observer. A second given point is the location of the sun. For example, on June 21 at 6pm the sun is 20° above due west. As a third point, you can, for example, define the position of the constellation Virgo. Virgo always moves on the celestial equator and is located above south three months after March 21 at 6pm, because it moves about one quarter per three months faster than the sun (comp. Fig. 7.4). The position of the ecliptic on June 21, 6pm solar time, is therefore the same as on March 21, or 12am sidereal time (see Fig. 7.3b).

After finding the standard positions of the ecliptic during the four seasons, you reach the insight that the ecliptic performs the same *pendulum motion* in 24 hours as over 365 days. The exact proof of this observation and the idea of the ecliptic ask us to envision the daily and yearly back and forth shuttle of the ecliptic in such detail that we can understand its conformity with a natural law and allows us to make predictions into the future. For example: In which season will the constellation Leo be above south at midnight?

This task can be solved as follows by using the complete Figure 7.1 as a blackboard drawing: As the position of the zodiac is given for March 21 at 6pm, we first have to rotate by six hours, that is 90°, in the same direction as the daily path advances. At 12 am, Leo is then positioned about 50° above south-southwest. As we want to have the lion directly above south, we have to turn back the zodiac by 30°, which is about a month. This brings us to the month of February.

This kind of tasks requires a lot of concentration and presence of mind from the eleventh-grader. To be sure, you will always have to create a consciousness of where you stand as an observer [consciousness of the own observational viewpoint].

The daily path of the sun—including the night portion of it—must be related to the corresponding signs of the zodiac, even if they are invisible during the day. Students therefore will need to develop not only trust in their concrete observations of the sky, but also in the structure of order developed by their own thinking. The external, sensually cognizable must be put in context with the internal—the steps of your own formation of justice.

I orient myself either towards "civil time," in which the sun repeats its path every 24 hours—but stars accomplish more than one revolution per 24 hours—or I use the practical advantage of the astronomical [sidereal] time, which starts a new convolution when the stars also start a new revolution [in 23 hours and 56 minutes]. This means that the student has to consciously walk away from his/her limited idea about time and receive in exchange an internal breakpoint at the steady return of the same star at the same place and the same sidereal time [astronomical clock].

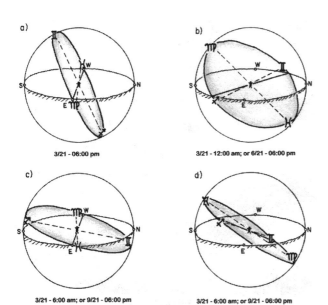

a) 3/21 - 06:00 pm

b) 3/21 - 12:00 am; or 6/21 - 06:00 pm

c) 3/21 - 6:00 am; or 9/21 - 06:00 pm

d) 3/21 - 6:00 am; or 9/21 - 06:00 pm

*Fig. 7.3:* Variation of the ecliptic position after 6 hours, or 3 months, respectively. Use Fig. 7.4 for a better understanding of the position of the signs of the zodiac. Drawing by Sarah Lewis Crow

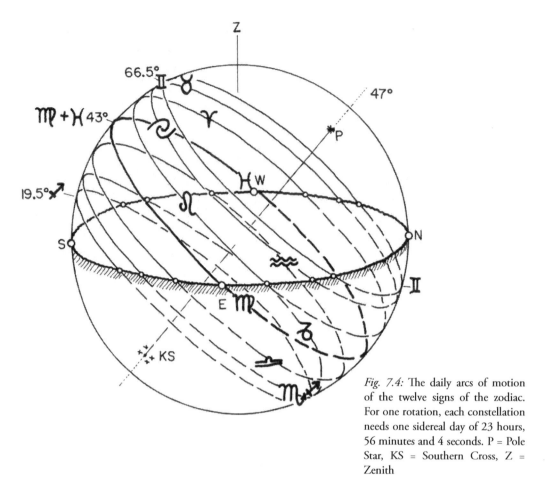

*Fig. 7.4:* The daily arcs of motion of the twelve signs of the zodiac. For one rotation, each constellation needs one sidereal day of 23 hours, 56 minutes and 4 seconds. P = Pole Star, KS = Southern Cross, Z = Zenith

A new particular *consciousness of time* has to be developed in addition to the consciousness of location.

In order to avoid premature confusion, it makes sense to prepare completely labeled drawings with special care [as well as board drawings in the main lesson book] and always become clear using these drawings. And allow the students to correct themselves if they misjudge something (see Figs. 7.1 and 7.4).

Let us get back once more to the option of stating predictions: If the context of motions in the sky developed due to observations that look back in time, then thinking can also look ahead into the future. This can be verified later through additional new observations. In this way the student can get to another important experience: In the cosmos, there is a certain order caused by the return of events, but nothing repeats itself exactly in the same way; boredom does not arise.

### The Platonic or Great Year

When you compare the position of the sun on March 21at 6am [vernal equinox] relative to the starry sky over the years, you realize a very slow movement of the starry sky past the rising sun. In the course of 72 years, a star will move visually 1° further to the left of the sun. In approximately 25,920 years, the *vernal equinox* moves backwards through the zodiac, and one revolution in the Platonic Year takes place.

The inner significance of the Platonic Year can be brought up easily with the students when talking about the three motions of the *Copernican worldview*. (Kuhn 1957) In order to describe the revolution of the earth around the sun, Copernicus identified three motions to each other: 1) the earth's rotation around itself, 2) the revolution of the earth around the sun, and 3) the rotational motion of the earth's axis, which is tilted by 23.5° away from the plumb line of the ecliptic, around the ecliptic plumb line, thus forming a conic surface around it. Let us assume that the earth axis is tilted to the outside when the earth displays the second motion. This would mean that it would always point to the outside as when the earth moves around the sun. Only the coordination between the second and third motions results in the approximate parallel orientation of the earth's axis, which causes the four seasons.

But now it is the case that the velocities of the second and third movements are not exactly the same. The relations between cosmic events are not simple integer multiples [incommensurability] (refer to Steiner 1921). Out of this situation the necessity of the Platonic Year arises as an inner cosmic law, because the rotation axis advances due to the third motion every year a little further and thus causes the shifting of the vernal equinox. This is a good example of the interference of rhythms of an incommensurable ratio, leading to a new rhythm.

## The Heliocentric Worldview

In addition to these more observational tasks, the students can exercise their imagination and look at the same motions from a heliocentric point of view. I have to watch myself—more or less from the outside, how I perform a complex, rhythmically repetitive motion together with my sight horizon, while the sun stands still in the center (comp. Fig. 7.5). Now I have to let the sight horizon move as complex as the ecliptic before, while now the ecliptic is the stationary plane, just the way it was the horizon before. For this fact it may have become clear that the conscious knowing how to handle the heliocentric approach requires much more self-confidence than standing within the geocentric system. It is now of crucial importance to help the student experience that none of the approaches is right or wrong but each of them must be treated consciously and consequently.

Misjudgments occur if I slip from one approach into the other without even noticing it, due to a lack of methodological awareness. It can be liberating if you experience in your consequent work that different Weltanschauungen [worldviews] lead to discovering reciprocal theories and predictions.

This is a beautiful example of the difference between the act of thinking, which is executed individually, and the content of thoughts, which are universal and always right. Different worldviews are justified only if they do not overstep other worldviews and do not mix with them. This thinking experience is almost a therapeutic necessity in today's world, where overstepping your competence [exact natural sciences versus life sciences] can be disastrous.

## Sun—Earth—Moon

Now we can add the moon for strengthening and enlarging on what has been practiced up to now. As one month—the duration of a block—is a controllable period of time, we can set for the students many observational tasks. These observational effects are now put into an inner context by alternatively but correctly applying the geo- and heliocentric worldviews. If you focus only on angle measures and angular motions up to now, you can consider focusing on a three-dimensionally correct perspective of the three-body problem Earth—Moon—Sun. This results in an understanding of the various types of eclipses. These phenomena provide another opportunity to recognize rhythm interferences and grasp them mathematically. The more simple eclipse periods are a good example (e.g., Saros cycles; compare to Freie Hochschule für Geisteswissenschaft Dornach; Held 1999; Bangert et al. 1991).

In this main lesson section that takes at least one week, there is again plenty of opportunity for exercises. Draw a certain moon phase at a certain part of the horizon and ask the students for the time of day and season when they could expect this picture as an observation (comp. Figs. 7.6 and 7.7).

The slim crescent moon in Figure 7.6 is positioned close to the point of sunrise. Therefore the time must be shortly before sunrise. The sun will follow the moon in about two hours and rise in the east. This means it must be autumn or spring. We find the answer when we recall how the ecliptic is positioned in the morning at 6am in spring or autumn. If the sun in the autumn is positioned in Virgo, the ecliptic at 6am in the morning is in a very steep orientation at the southern sky. The answer is therefore: autumn at 4am in the morning.

The daily path of the moon in Figure 7.7 is very flat. The moon can be seen in the sky for only about eight hours; almost full illumination suggests that the sun is setting. The full moon would rise above southeast in summer, the waning half moon in spring. Our picture therefore shows the situation in May in the early evening.

When studying the moon, it suggests itself to study the obvious or hidden effects of the moon on life on earth. The subtle country sayings on moon phases on the one side and weather cycles and plant growth on the other side are a source of information. Reproductive cycles in animals and human beings point to further contexts (see Endres and Schad 2002). When talking about the lunar nodes—without them, we could not understand eclipses—we can suggestively point to biographic rhythms in the human life. As you see, we get much closer to the concept of intentionality when studying the moon quality.

The more time you dedicate to the study of the moon, the more opportunity you will have to portray very subtly the processes on earth and observe your own soul life. If you let students express moon experiences in the form of a written essay, their soulful dismay will show very clearly and creatively.

### Earth—Stars—Planets

In a third phase of study we can discuss a few planets. The relation between wandering stars [planets] and fixed stars, the phenomenon of *planetary loops*, and the alternation between forward and retrograde motion—all can be recognized and constructed heliocentrically as well as geocentrically (compare to Haag 1998). A new field of rhythms with a certain number ratio appears. The same way you developed eclipses in the previous section of the block, you are now including the peculiarities of conjunctions, oppositions and quadratures into the discussion.

If you are careful enough to include in the discussion the moon quality and moon effects on earth, corresponding questions will arise during the discussion of the wandering stars. What is the qualitative difference between the brilliance of Venus compared to Mars? What could be the earthly counterpart to the four-month rhythm of Mercury, in terms of conjunction and opposition? A resource

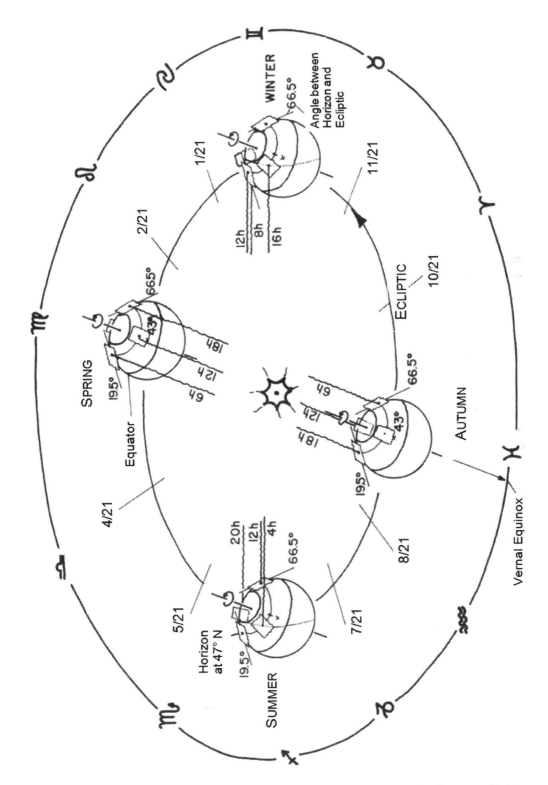

*Fig. 7.5:* The earth's movement as seen from a heliocentric perspective far in outer space (not drawn to scale). Note the daily [diurnal] and yearly [annual] shuttle of the observer's horizon at 47° N [here in Switzerland]. The angular values close to the small rectangles of horizon show the corresponding angles between the moving horizon and the fixed plane of the ecliptic. The hour values represent the mean time for sunrise and sunset at 47° N.

for teachers on these topics is the important work *Plant and Cosmos* by Kranich (1997).

Toward the end of this section of study, the question of the exceptional quality of the earth in the solar system is asked [e.g., temperature conditions on earth in regards to the physical states of water and $CO_2$]. Characteristic descriptive fields develop when summing up the quality of the motion of a planet, its physical characteristics [temperature, atmosphere, rocks, etc.], and your feelings of how planets are so-to-say special cases. The summary of and balance between such special cases can result in the description of the earth. In this way, the exceptional quality of the earth within the cosmos can be appreciated in astonishment. Recent literature about the physicochemical characteristics of the planets can be found in the book *Planets, Wanderers in Space* (Lang and Whitney 1991) and the anthology *Planets and Their Moons* (Wielen 1997).

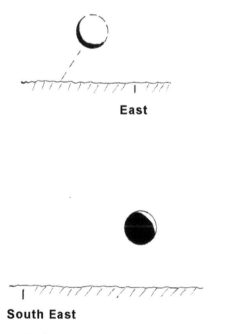

**East**

**South East**

*Figs. 7.6 and 7.7:* Two moon positions with indication of the compass point on the horizon. The time of day and season of these positions can be found through careful reflection [exercise].

## Rhythm of the Ice Ages

In a final fourth week of the block, the class is ready to grasp the astronomical relation between cosmos and earth. Three rhythms are identified and related to each other. First is the rhythm of the *Platonic Year* [approximately 25,920 years]. Secondly, gained from the discussion of Kepler's laws, is the rhythm of alternation between the perihelion and aphelion, the rhythm of the eccentricity of the earth's orbit [approximately 110,000 years]. The third rhythm is newly noticed. It is the cyclic modification of the earth axis tilt relatively to the ecliptic. The alternation between 21.0° and 24.5° and back takes about 41,000 years. If you interlink the three rhythms, you receive the periodic alternation of the irradiance of the sun on earth (comp. Fig. 7.8).

In the Northern hemisphere, periods of thousands of years of mild winters combine with cool summers. This facilitates an increase of glaciers in the Northern hemisphere. The following period of cold winters and hot summers is unfavorable for glaciations. In cold winters there is little precipitation, causing more water to melt during hot summers than what was deposited during winters. If you now hope to draw a causal linkage between cosmic rhythms and glaciations cycles, you will be disappointed as the cosmic rhythms release just either a favoring or inhibitory effect. (Jouzel et al. 2007) The morphological configuration of the earth and the living conditions on earth constitute another condition that may enable a glacial or an interglacial period. The publication *Plateau Uplift and Climatic-Change* (Ruddiman and Kutzbach 1991) provides information on the influence of the high plateau of Tibet on the formation of ice ages. Favored by the high plateau, a decrease in temperature causes a large, year-round snow cover, which boosts the chilling effect by reflecting the cold sunlight [albedo effect]. Another condition is the vitality of the plankton in the oceans. Plankton density increases or decreases in relation with changes in ocean currents—caused by the calve of large glaciers, which lead to a modified $CO_2$-content of the atmosphere and by this to temperature changes (see *What Drives Glacial Cycles* by Broecker and Denton 1990).

For example, dependent on the North and South Americas being connected by a land bridge in Central America, the surface currents in Central America [Gulf Stream] and the central eastern Pacific were completely different, causing in return the heat transfer by flowing water to change drastically. (Blankenburg 1999)

Thanks to current scientific research, the following interpretation is justified: Cosmic rhythms provide impulses, which are picked up by the earth in its constellation and its life processes like an organ within a superior organism [cosmos]. In other cases, one of these impulses might also be missed. The discussion of the ice age cycles becomes meaningful when you indicate what multiple alternations between warm and cold times might mean for the earth. These alternations are in return impulse and condition for a refreshing evolution of life on earth. The receding and advancing of the glaciers stimulated large scale migrations of plants, animals and, last but not least, the predecessors of *Homo sapiens*.

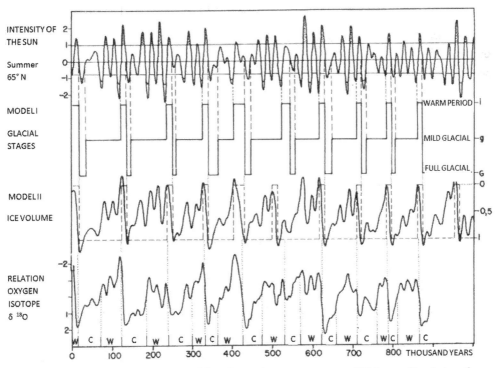

*Fig. 7.8:* The upper panel shows the variation of the solar irradiance in summer at 65° N, caused by the interference of three cosmic rhythms: Platonic Year, rotation of the apse line [eccentricity], and fluctuation of the tilt of the earth's axis relative to the ecliptic plane. The lower curve shows the variation in the $^{16}O/^{18}O$ isotope ratio, as measured in ocean sediments. This fluctuation corresponds with the growth and reduction of the ice shields. The two intermediate panels show the result of a modeling calculation in two approximation of the variation in the ice volume. (from Paillard 1998)

This keyword of developmental impulses is an appropriate and valuable conclusion to this block. The new questions arising are put aside, to be picked up and processed in the earth science block of grade 12.

## On the History of Astronomy

The reader might ask when the historic elements of astronomy will be covered. It is indeed a good possibility to sidetrack into the history of science when it is appropriate. The most obvious example is the change from the geocentric to the heliocentric worldview as a truly revolutionary deed of Copernicus and Tycho Brahe. For this subject, *The Copernican Revolution* (Kuhn 1957), is a valuable resource. As a next step, you could honor Johannes Kepler, who successfully tried the mathematical approach. These can be followed by the celestial mechanic Isaac Newton and furthered to the modern telescope astronomers. In these discussions it is important to show how the deeds of knowledge of individual personalities affected the cultural development but with a certain time lag. Cultural development is experienced as a step-by-step increasing in the self-consciousness of the human being. These steps are also being taken by the young person but he/she can recognize and study them more easily and consciously through the chronology of historical events.

## Relations to Mathematics and Surveying

One point of connection to the surveying in grade 10 is the use of the sextant. Using the sextant is like watching from the sky where exactly on earth I am standing while sailing the wide sea. (Bauer 1992; Budlong 1978) Again you find the phenomenon of distancing and opposing yourself as a necessary precondition for observation. Now the maximum distancing is reached: In order to determine my location, my imagination needs to move out into the cosmos. In this way the most important step to understanding satellite navigation is taken. If spherical trigonometry has already been covered in mathematics, you can set beautiful and continuing exercises for the students. As a connection between the use of the theodolite and the practice in spherical trigonometry, you may use the book *Astronomical Determination of Location, Time and Azimuth with the Kern DKM3-A.* (Müller 1971)

## Summary

From the weighing of the subjects to be covered, hopefully it has become clear that it is—if at all possible—the task of the earth science teacher to teach this block. The argument that an earth science teacher would not be competent enough to teach the material is unfounded. You can read in the third scientific lecture cycle (Steiner 1921) that a natural scientist who does not deal enthusiastically with astronomy may not be a Waldorf high school teacher for long. It is also possible to phrase this claim positively: The area of astronomy is that exciting, and the observational options in the sky so inviting, that an examination of it can only be profitable. I was lucky enough to experience how 11th graders went through the astronomy block with increasing fascination. They realize how much can be studied above the earth and what kind of satisfaction occurs when a new world context is created in their own thinking.

Even if the presented draft makes distinct claims, I hope that they will not miss its target, namely to encourage to get this important block at Waldorf schools generally accepted.

# Chapter 8

# The Economy of Energy in the Eleventh Grade

## Introduction

To learn about the situation of the 11th grader in relation to the study of human science and the appropriate pedagogic tasks for this developmental age, please refer to the descriptions in chapters 1 and 7. The second possible block for this grade should be a crafts block carried by a strong life skills element. A teaching unit should be understood as a crafts block if it is taught in three double periods per week for up to 4–6 weeks. These double units take place either from 10 am to 12 pm—or in the afternoon. [In contrast to this, main lesson blocks in Germany and Switzerland are usually taught from 8am until 10am—tr.] The crafts schedule alternates blocks in craft skills, arts and technology. While the astronomy block should exercise the thought process of intentionality, the crafts block on the economy of energy should offer exemplary insights into one key site of modern civilization. This allows for covering different aspects to different levels of depth as needed. For one it is the point to rediscover previously covered physical principles in technical applications and deepen this rediscovery through exercises. Sociological and life science issues are addressed when focusing on the working conditions in mining, refining and conversion of energy resources.

Economical and political considerations are added when producing and consuming regions are covered. At this point, the topic of a responsible handling of the world by the human being is inevitable: The themes of ecology become most important. When arriving at this point and realizing that the existing large-scale technologies of energy conversion are a cul-de-sac, the relieving step towards alternative technologies is likewise a step towards reinventing the quality of living—which leads back to sociology and life science. When discussing possible implementations of alternative energy concepts, the interrelations between ecology and politics are described in their basics.

In this sketch of what can come to life in such a block, it is easy to see that a strict question and answer schema of cause and effect is not useful. Likewise, thinking and imagining circulation processes alone does not do justice to this problem. Instead of this, students experience an example of intentionally interrelated events (comp. chapter 7). The uses of certain technologies influence and motivate human creativity. Vice versa, the inventive genius of the human being originates new technologies that in return influence our way of life.

As this is a second block and is taught after main lesson, it is conducted in a different instructive mode. Instead of a classic main lesson book, a workshop journal is kept, developed mainly during class time. For practicing this work technique, this block adds dealing with statistics and learning how to read challenging journal articles. In class discussions students learn how to present and support their own positions and judgments.

## Exemplary Flashlights on the Block Content

### About the Concept of Energy

Although the literature speaks almost exclusively of energy production and energy consumption, we will discuss the physical law, according to which there is only energy conversion. Energy therefore has the ability to perform work. If you look at various transformations, you will realize that the efficiency of the transformation into light, heat or driving force never quite reaches 100%. A large proportion transforms into mostly not actively "used" waste heat. From this perception, we introduce the concepts of primary energy, secondary energy, end-point energy and useful energy. The technical literature defines primary energy as the so-called raw material [coal, crude oil, natural gas, uranium, wood, water in an elevated storage, garbage] from which usable energy forms originate after the first transformation. Secondary energy has already gone through a transformation. Examples are electricity from water or nuclear power, or gas from coal. The form of energy that is transformed in a consumer household, in a factory or during a technical locomotion is called end-point energy. Useful energy [usable] is finally the one form of energy, which serves consumers in our civilized way of life after being transformed into light, heat, power or chemical energy bound in batteries.

Astonishing results can be gained from a Swiss energy review, which is provided to the students as a statistical summary. (Schweiz. Bundesamt für Energie and Schweiz. Bundesamt für Energiewirtschaft; similar results for most other countries, e.g., European Commission: Directorate-General for Energy and Transport; United States: Energy Information Administration) Taking 100% primary energy in 1993 as our starting point, end-point energy amounts to 75%; the energy loss of 25% is caused mostly by the five Swiss nuclear power plants, which reach an efficiency of only 32 to 37%. Useful energy—which is finally usable for the consumer—amounts to only 43%, because 43% of the end-point energy is lost in energy conversion. Altogether this means that 57% of the energy is lost on the way from primary energy to useful energy, and the efficiency of the energy conversion devices is very low (comp. Fig. 8.1). The frontrunners in low efficiency are nuclear power plant at the transition to the end-point consumption, the gasoline combustion engine and the conventional light bulb on the level of energy conversion at the consumer site.

### Embodied Energy and Harvest Factor

If you want to judge the efficiency of an energy conversion event properly, you have to account for embodied energy. Embodied energy is the conversion of energy which happens during the production of a device or providing a service. A 220-liter refrigerator converts about 450 kilowatt hours [kWh] of energy per year. For its manufacturing, about 1400 kWh of embodied energy is used. The calculation of the yearly consumption of embodied energy for an average Swiss citizen results in the astonishingly high number of 30,000 kWh per year. Compared to this, a family of four with a medium standard of living [3-bedroom apartment, medium-sized car driving approximately 15,000 km = 9320 miles per year] uses only slightly more energy at 35,000 kWh per year. This number is double if the family travels by plane once a year to the United States [or from the US to Switzerland, according to Strahm 1992]! Even more impressive is the example of a flashlight battery, with a usable energy conversion of only 0.012 kWh, while the expenditure in embodied energy amounts to 0.65 kWh; so calculating the efficiency including the embodied energy, the flashlight battery reaches only 1.8%!

According to the May and June issues of *Energietechnik* (1996), the harvest factor is introduced in order to describe the ratio between usable energy and embodied energy (comp. TA 3.5.1996). In an energy conversion device, the harvest factor represents the usable energy [room temperature, power] produced during its lifetime divided by the expenditure in non-renewable energy resources for the manufacturing and disposal of the device as well as for the processing of the energy carrier [oil, gas, coal, uranium, wood]. Positive energy balances with harvest factors above one are reached by modern wood heaters

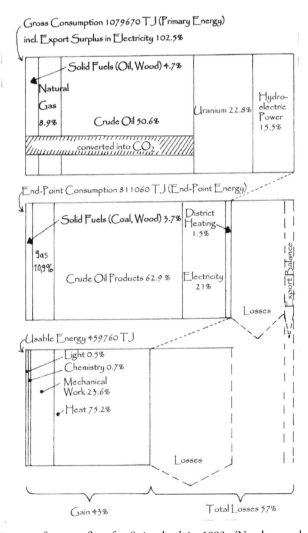

*Fig. 8.1:* Simplified diagram of energy flow for Switzerland in 1993. (Numbers and schema according to Schweizerische Gesamtenergiestatistik [Swiss Energy Summary] 1994)

[7.1], solar warm water heaters [4.0] and solar electric generators [1.6]. Newer solar collector systems with a life expectancy of 20 years reach a harvest factor of 11.2, solar cell systems with a life expectancy of 30 years reach 6.2. (Infoenergie 1994) All other devices show a negative energy balance: heat pumps [0.8 using Swiss electricity, 0.6 using EU electricity], heating systems using gas or oil [0.7], coal heating [0.5], electricity conversion originating from Swiss power plants [0.3; with 40% nuclear power plants, 25% hydroelectric power plants and 35% storage power plants], electricity conversion originating from EU power plants [0.2]. These harvest factors are published in the study *Graue Energie und Umweltbelastungen von Heizungssystemen* [*Embodied Energy and Environmental Impact of Heating Systems*], May 1996, which is a condensed version of the study Ökoinventare für Energiesysteme [Eco Inventories for Energy Systems] by the ETH Zürich and the Paul Scherrer-Institute, Dones 2003; Frischknecht et al. 1995.

## Non-Renewable Energy

### The Formation of Raw Materials

To begin, we will discuss the geological origin of the non-renewable energy sources of crude oil and coal (refer to Fig. 8.2). Corresponding details would also be valid for natural gas. All of these energies originate from a massive formation of biomass, caused by photosynthetic processes, which split carbon dioxide into carbon compounds and oxygen. These processes lead to a reduction of the carbon dioxide content in water and atmosphere and consequently a reduction of the greenhouse effect, which means a cooling of the earth. (Graedel and Crutzen 1989, 1993, 1995; Keeling 2006; comp. chapter 9)

It is something special that the carbon compounds in crude oil and hard coal were deposited in depths of several kilometers in such a way that they could not bond to atmospheric oxygen (refer to Fig. 8.2). The events described above—the enormous presence of life and successive die-offs leading to carbon deposits—were the preconditions for the development of modern life on earth with ideal global temperatures and, more importantly, the temperature conditions for a rhythmic sequence of ice and warm ages.

This process peaked first during the Carboniferous (350 to 290 mya, or million years ago; refer to Fig. 8.3) [In the US the Carboniferous is usually divided into two periods, an earlier period named Mississippian and a later period referred to as Pennsylvanian—tr.]. The associated cooling led to a mild ice age and important developments of life on land [solid ground]. During the major extinction of life at the end of the Permian, the mean annual temperature on earth had risen to approximately 22°C. The next cooling period was accompanied by a reduction in carbon dioxide during the Jurassic [160 to 130 mya], and even stronger during the Cretaceous [130 to 60 mya], during which mostly crude oil was formed (refer to Fig. 8.4). This way a global temperature developed which provided the basis for the rhythmically appearing ice- and warmth-ages.

If man opens these carbon deposits and uses them for fuel, it is a logical consequence that the $CO_2$ content in the atmosphere will increase and global temperature will rise to values corresponding to the Mesozoic. The dangerous difference of this reversed process induced by human activity is its 100,000 times faster speed. Life cannot adapt to this process or actively shape it. Nature

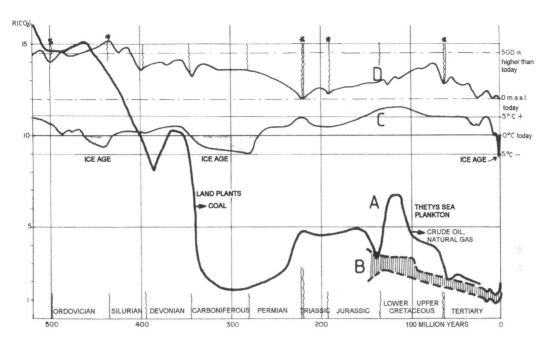

*Fig. 8.2:* Climate and carbon dioxide content between the Cambrian and today. A shows estimated carbon dioxide concentrations from seafloor spreading and land covering data. B shows the field of possible carbon dioxide concentrations gained from carbon isotope ratios. R ($CO_2$) indicates the ratio between historic and current (= 1) concentrations. C, from paleoecological data, shows the fluctuations in global temperatures over the course of time, shown in contrast to today's climate. D shows the fluctuations of the sea level, starting at today's reference 0 m.a.s.l. Asterisks (*) indicate major extinctions of marine life forms. (A, B, D drawn according to Graedel and Crutzen 1995; C drawn according to Allègre and Schneider 1994, supplemented)

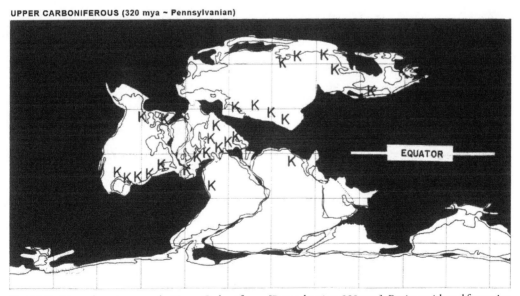

*Fig. 8.3:* Position of continents at the Upper Carboniferous [Pennsylvanian, 320 mya]. Regions with coal formation marked with a K. The equatorial tropical belt, the coastal subtropical coal regions and the coal area of the northern temperate forest zone of today's Siberia are clearly visible. (According to Closs, Giese, and Jacobshagen 1987; Smith, Hurley, and Briden 1981, supplemented)

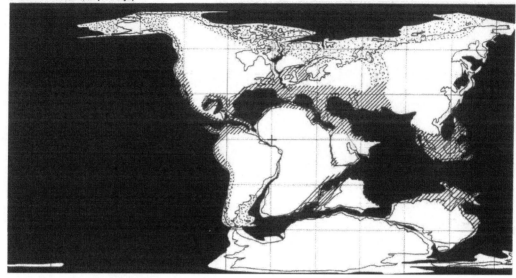

*Fig. 8.4:* Position of the continents at the beginning of the Cretaceous. Continental areas covered by the shallow waters [continental shelf] of the global Tethys Sea are indicated with beveled shading. This is where most of the crude oil and natural gas deposits have formed. (According to Smith, Hurley, and Briden 1981; Stanley 2008; Stanley and Schweizer 1994, supplemented)

experiences this reverse process as a shock, although we humans may think this process as slow, as it has progressed over several generations. A logical consequence of these reflections is: Non-renewable energies such as coal, oil and gas may be taken from earth only at such a rate and speed as the rate and speed needed to form corresponding amounts of these carbon compounds. This would mean that the extraction of non-renewable energy would have to be reduced by several 100,000 times, which would be at a rate sufficient to produce synthetic substances from these resources but not to burn them.

### About Hard Coal Mining

For example, coal mining has required the development of highly qualified workers and a highly developed professional ethic. Mining requires intensive safety measures and a permanent monitoring of all ducts and tunnels. The miner experiences every day how the mountain changes dynamically in its depth. Human engineering tries to meet this deformation activity consequently and without remorse. In this line of work in the depth, it is completely natural that nobody works alone; hour for hour the workers are watching each other to avoid possible accidents.

If it were not for government financial support, these extreme efforts in qualified labor would cause coal prices to be four times higher than oil prices, and, in contrast to oil production, at these prices there would not even be profit. It is special for mining jobs that the workers develop important faculties: considerate cooperation, reverence for the overwhelming forces of nature, forward-looking craft skills and interpersonal solidarity. When looking at this economic sector in Europe, it is obvious that coal mining is economically sentenced to death, and that state aid only delays the eventual closing down of mines.

## About Crude Oil Extraction

In contrast to coal mining, crude oil extraction is much less labor-intensive. True, there are huge high-tech expenses and the payroll of utilizing competent scientists to discover natural oil traps and determine the location of drill derricks, which is in principle similar to coal mining. The oil production itself is simple. Once an oil well is opened, the outflow of the oil under pressure only needs to be controlled to allow its smooth transportation via pipelines. No worker has to reach the oil well itself. This explains why oil prices are still low despite tremendous profits of land owners and the oil-processing industrial complexes. It is justified to say that oil production is the cleverest and most efficient exploitation of any natural resource!

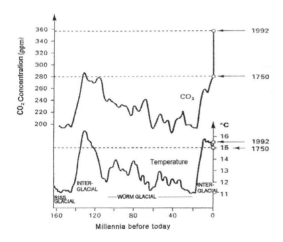

*Fig. 8.5a:* CO$_2$ concentrations and global temperatures within the last 160,000 years. Measurements are based on ice cores from the Vostok station in Antarctica. The increase in CO$_2$ over the last 250 years is completely outside of the cyclic norm. (from Houghton and Woodwell 1989) [The Riss glacial is roughly contemporaneous to the Illinoian glacial in the US, the Würm glacial to the Wisconsin stage—tr.]

## How Limited are Non-renewable Energy Resources?

An assessment of the supplies is not a simple task. From the previous representation it should be clear that especially crude oil and natural gas resources will run out soon. It is relatively clear in which earth regions to find oil and gas deposits, but weighing costs and benefits to exploit known deposits is much more uncertain.

When the classical oil deposits are leeched or pumped empty, the oil refining industry will start to exploit oil shale. The extraction has already begun as open-pit mining and has resulted in massive landscape changes. It is difficult to estimate the accompanying technical problems and political and ecological conflicts. In 1993, oil reserves globally [without oil shale] amounted to 136,000 million tons. (Fischer Weltalmanach 1995) During the last several years, exploitation increased 1.4% per year—and, if it stays like this, oil deposits will be completely exhausted at the latest in 43 years (comp. Fig. 8.5b). The political dimension of this situation is even more dramatic when we consider that only 6% of the reserves are located within the industrial regions of North America and Europe. Another 8% of oil is stored in Russia and China. The remaining 86% is in Latin America, North Africa and the Middle East. In other words, if production stays at the current level in all regions of the world, deposits in Europe and North America will be exhausted around the turn of the millenium, leaving the entire industrialized world dependent on the oil resources in Latin America, Africa and most of all the crisis-ridden Middle East. Saudi-Arabia, Iraq, Kuwait, the United Arab Emirates and Iran alone have a reserve potential of 87,000 million tons of crude oil which converts to an exploitation potential

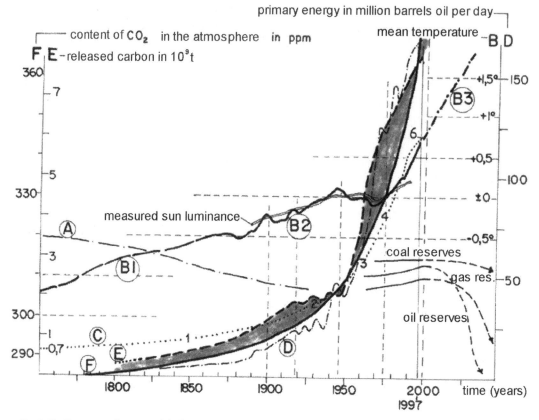

primary energy in million barrels oil per day

content of $CO_2$ in the atmosphere in ppm    mean temperature

F, E – released carbon in $10^9$ t

*Fig. 8.5b:* Connection between global temperatures, carbon dioxide content and energy conversions between 1750 and 1995 and forecasted values until 2050. A shows the trend towards temperature changes based on astronomical facts but without any human influence. B shows the temperature graph; Bl was gained from historical data, B2 is based on measurements, B3 shows an average forecast. C shows the population increase in billions of people. D represents the expenses of energy conversions [primary energy], measured in Mio. Barrels of oil per day. Graph E, which is almost parallel, shows the annual release of carbon originating from the combustion of fossil fuels [oil, gas, coal, wood] and changed land use such as forest fires. F documents the atmospheric $CO_2$ content; compared to graph E, this graph shows a delayed increase. At the right border are graphs showing estimates for the remaining reserves of oil, gas and coal (D and estimated until 2050 according to the Statistical Review of World Energy, British Petroleum Company (1992–2009) and the Fischer Weltalmanach, Redaktion Weltalmanach 2009; comparable to The World Factbook (United States, Central Intelligence Agency); B2, B3, F according to Graedel and Crutzen 1995; B between 1880 and 1990 according to Hansen and Lebedeff 1988; E according to Houghton and Woodwell 1989; B2, B3 according to Jones and Wigley 1990.

at their disposal of more than 100 years (numbers from Fischer Weltalmanach 1992, and BP Statistical Review of World Energy 1993).

When keeping these potential dependencies in mind, it is all of a sudden easy to understand why the Russians in Chechnya, the Americans and Europeans in Iraq and in the living area of the Kurds, and Palestinians are involved in military actions, which on a first view seem politically unwise and ethically irresponsible. They are, however, logical and consistent if it is all about gaining or securing political influence on these crude oil-producing regions as well as the areas through which the critical pipelines are drawn.

## The Non-cycle of the Uranium Technology

The dispute about the development and use of civil and military nuclear technology is stuck; even eleventh graders have quite solid opinions. Despite of this, it is still possible to discuss one specific aspect: Compared to the geological cycles of carbon in the use of coal, oil or gas, or in the closed loop of a hundred-year cycle of wood use and forestry, no cycle is possible in the case of nuclear power. Uranium ore, the necessary raw material for the operation, was formed very early in the earth's history. Uranium oxide was formed during the late Precambrian as a marine precipitation in connection with the process of photosynthesis performed by early protists (comp. chapter 9). The release of oxygen into the water led to the formation of uranium oxide, which was subsequently deposited. Again it is the reduction of the atmospheric $CO_2$ content and the reduction in the earth's surface temperature that brought about the formation of uranium oxide.

Since its formation, uranium ore has gone through several metamorphoses and is predominantly mined in surface mining causing massive *waste dumps* that quickly emit most of their radioactivity into the environment. Before mining was begun, uranium ore deposits were geologically screened off from the environment and only very little radioactivity was released. Nowadays spoil dumps are usually not treated in any way. A diagram of the mass

| Region | | oil production in mill. tons 1989 | oil reserves in bill. tons 1989 |
|---|---|---|---|
| TETHYS SEA | Saudi Arabia | 225 | 35 |
| | Iraq | 138 | 13 ⎱ gulf war ! |
| | Kuwait | 91 | 13 ⎰ |
| | United Arab Emirates | 91 | 13 |
| | Iran | 145 | 13 |
| | Lybia | 53 | 3 |
| | Algeria | 45 | 2 |
| | Middle East Total → 850 | | → 92 |
| | Venezuela | 96 | 8 |
| | Mexico | 144 | 8 |
| | USA | 426 | 2 — gulf war ! |
| | Indonesia | 60 | 1 |
| | Tethys without Middle East 426 | | → 20 |
| NORTH PROVINCES | PR China | 136 | 3 – ! |
| | USSR | 608 | 8 ⎱ |
| | West Europe | 120 | 2 ⎰ gulf war ! |
| | Canada | 93 | 1 |
| | remaining regions | 540 | 10 |
| | North provinces total → 1'497 | | → 24 |
| WORLD | total | 2'773 | 136 |

| selected regions | Oil Production in mill. tons 1989 | oil conversion in mill. tons 1989 |
|---|---|---|
| USA | 426 | 900 |
| West Europe | 120 | 430 |
| Japan | 5 | 230 |

*Fig. 8.6:* Global oil production, oil conversion and oil reserves for 1989. Crude oil contributes 40% to the total global energy conversion. Kuwait alone has as many oil reserves as the US, Russia, western Europe and Canada combined! (from Fischer Weltalmanach 1992; assembled by students)

balance for a single year of operation of a 1000-megawatts [MW] nuclear power plant helps to illustrate the dimensions of this problem (Fig. 8.7).

The extraction of 440,000 metric tons of uranium ore is needed to convert process heat into electrical power in such a nuclear power plant (of the Gösgen or Leibstadt type, both in Switzerland). [The two blocks of Three Mile Island near Harrisburg, PA had/have a comparable performance at 875 and 906 MW—tr.]. From these are gained 33 metric tons of uranium fuel. The remaining approximately 400,000 t of rock material, containing decay products of uranium such as thorium, radium and lead, is dumped onto the soil dump. During the production of uranium gas from the ore, which is suitable for the enrichment of the fuel rods, additional waste is generated. From the original radioactivity of 10 Peta-Becquerel in the rocks, only 0.43 Peta-Becquerel can be used for enrichment, which means that 95% of radioactivity stays in the environment. The process of radioactive decay takes a very long time; the quantitatively important nuclide thorium 230 has a half-life of 75,000 years.

The consequences of radioactive decay are therefore not apparent immediately but are spread over several hundred thousand years. The highest radiation hazard occurs during uranium mining. In most cases, the mining sites are far away from the sites of nuclear power plants and therefore from the people who profit from cheap nuclear energy, whereas the mining areas are typically in the living areas of indigenous people such as reservations for Native Americans in North America, aborigines reservations in Australia and African tribal lands in South Africa. Radioactivity is also released in enrichment plants, the next step from uranium ore to fuel rods. As these facilities are often located in Third World countries and not in industrial countries, the release of radiation is often not supervised due to the lack of appropriate laws and monitoring facilities.

The operation of nuclear power plants in industrialized countries with high safety requirements is the least harmful part of the further path converting enriched uranium ore into heat and radioactive waste products. Spent fuel rods constitute a high risk potential. Up to now, and most probably for the next decades, an ultimate waste disposal of such fuel rods is not possible if the current safety requirements are taken seriously. Detouring these spent fuel rods via reprocessing plants finds increasingly less support due to the unsafe operation of these plants and the amount of released radioactivity. It is important to state at this point that nuclear fission releases products that never before occurred in the surface area in the whole history of earth. Due to the radioactive isotope decay series, new intermediates occur that will impact the biosphere for a very long time. The mode of impact and, more importantly, the long-term consequences are largely unknown. The path of uranium containing rocks to uranium ore, enriched fuel rods, spent fuel rods and decay products of reprocessing plants and further, to the continuously changing waste products, constitutes an *irreversible* process, which is the complete opposite of cyclic processes.

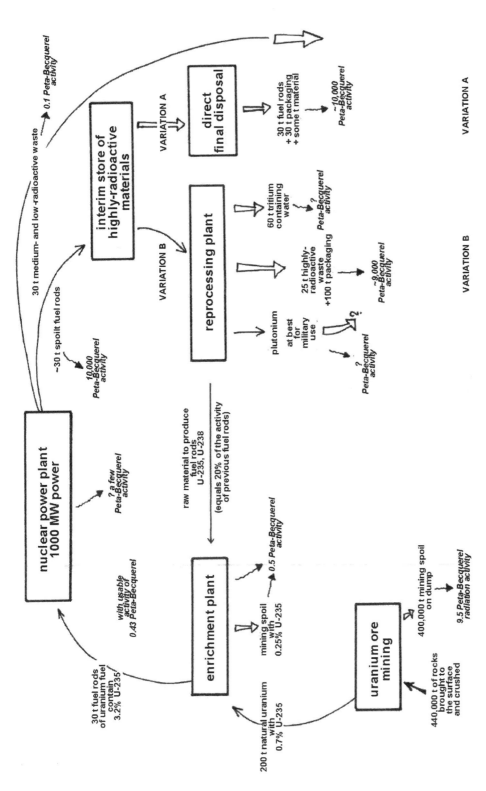

*Fig. 8.7* Materials and activity balance for the yearly operations of a modern nuclear power plant producing 1000 Megawatts of power. Peta-Becquerel are used as a measure of radioactivity; one Peta-Becquerel equals 1015 decays per second or a million times a billion decays per second. Simple arrows represent pathways of materials, double arrows show deposited materials, and waved arrows indicate released radioactivity. Another option of "burning" plutonium in fast breeder reactors is not shown, as this technology had not been successfully operated anywhere in the world. (Herrmann 1983; Herrmann and Knipping 1993, supplemented)

129

## Renewable Energy Sources

### The Cycle of $CO_2$ in the Use of Wood

The term "renewable energy" describes energy sources of which conversion does not lead to a reduction of natural resources. This includes all biomass, which is connected to care and conservation of nature. We will have a closer look at wood as an example. Biomass is formed during photosynthesis, or transforming carbon dioxide into glucose and releasing oxygen as a waste product. If non-harvested wood rots in a forest, bacterial decomposition transforms oxygen back into carbon dioxide. In principle the same thing happens when wood is cut and burned to produce heat (Fig. 8.8).

If a forest with a diversity of species is well taken care of and kept young by calculated wood cutting, the production of heat in wood combustion is ecologically sound. The cycle of carbon dioxide is a closed loop, meaning that the content of atmospheric carbon dioxide does not increase due to wood combustion. The only effect is a slight increase in the cycle of carbon dioxide formation and decomposition. The benefits for the living nature and the cultural world are further increased as well-kept forests qualitatively improve pedogenesis [soil formation] and the water balance of landscapes. The discovery that this is not an example of nature exploitation is very important for the students. First it takes efforts to cultivate and care for nature. Only then is the human being allowed to harvest wood from the forest in such a way that growth of the rejuvenated forest is improved. This energy conversion concept becomes even more convincing if we look at the qualitative side. Correctly cared for forests become high quality recreational areas. In addition, wood fueling counteracts the increasing alienation of mankind from nature. There might be objections that wood combustion leads to an unpleasant release of hydrocarbon compounds and nitrogen oxide pollutants. The new generation of wood-burning stoves achieves almost emission-free combustion, thanks to electronically-controlled wood gasification and combustion at high temperatures. Generated fumes as well as distillation gases are combusted using a lambda probe controlled afterburner. An oft-asserted objection is that there is not enough forest. However, just in Switzerland alone, wood currently laying fallow could provide up to 10% of the current heating requirements. With a sustainable and

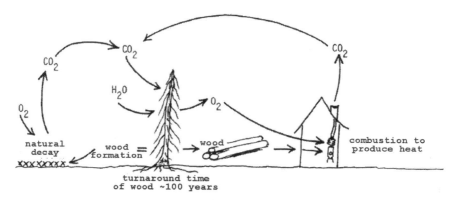

*Fig. 8.8:* The $CO_2$ cycle in connection to forest care and wood usage for conversion into heat. Within the growth period of a hundred years, a closed loop develops! If the forest is artificially kept young, a slight shortage in $CO_2$ develops, which can, however, be balanced by a subsequent stronger decay.

intensive care of forests, this percentage could be higher. In contrast to this potential, wood used as a form of energy amounted to only 2% in 1992.

## Solar Warm Water Generation

Warm water generation with the help of the sun is a good example for discussing decentralized energy conversion devices. These devices are technically easy to understand and can be built from scratch. No carbon dioxide is produced in their operation, and no raw materials are consumed. We will now describe the principle of such a device (Fig. 8.10). The centerpiece is a solar collector using the greenhouse effect (Fig. 8.9). Conventional flat-plate collectors consist mainly of a transparent glass or plastic plate, an absorption plate made out of steel, copper, aluminum or plastic, and a good insulation from mineral fiber or plastic foam. The absorber is painted dark and contains a fine tube system, in which the heat carrier, an antifreeze, circulates. When the absorber receives diffuse or direct sunlight, the light passes through the glass plate and is converted into infrared or thermal radiation. This is partially trapped inside the collector, as thermal radiation cannot easily pass back to the outside through the glass plate. This heats up the circulating liquid within the absorber, which is then moved by a pump in a closed loop to a heat exchanger. The heat exchanger is situated inside a hot-water tank.

Instead of using a pump, circulation can also be achieved by using the thermosiphon effect. If the collector plate is situated at the lowest level, the heated water rises and thus moves to the heat exchanger, from where it falls back to the collectors after cooling off. The brilliant feature of this solution is the self-regulation of circulation. Additional installations will require only a swing check

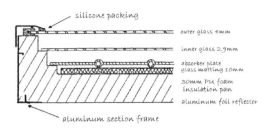

*Fig. 8.9:* Cross-section through a solar collector. (Rüesch, type Monoblock 4) The heat transfer medium flows through cavities within the absorber plate. (Schmutz, H. 1987)

valve in order to avoid a reflux of the stored heat during night. As security measures, a pressure relief valve and an expansion tank are included into the closed heat transfer fluid loop.

If a circulating pump is used, a small electronic control and two heat sensors will also be needed. The pump will be activated as soon as the heat sensor detects a temperature 5°C higher compared to the sensor in the lower portion of the heat accumulator. To be independent from the supply network, the needed electricity can be gained by adding a solar panel.

## Are Solar Systems Cost-efficient?

The following specifications are quoted from *Solare Warmwassererzeugung* [solar warm water generation], which was collected as an incentive program by the Swiss Federal Department of Economy (Schweiz. Bundesamt für Konjunkturfragen. Impulsprogramm PACER 1993 and 1995), as well as from a brochure by the solar energy information and advisory service of the Swiss committee for the use of solar energy. (Infosolar 1991)

First, is the solar irradiance in Switzerland [compares to approximately 45°N—or S in other regions of the world—tr.] sufficient for an economical operation of a solar system?

In the Swiss Midlands [46–48°N] global irradiance at the earth's surface ranges from 170 kWh per square meter in July to 25 kWh per m² in December. Global irradiance is the sum of approximately 45% direct radiation and approximately 55% diffuse radiation. Most residential areas in Switzerland reach an average global irradiance of 1 kWh per m² collector area. At an average sunshine duration of 100 h per month, this results per year in an average of 1200 kWh per m². Given an efficiency of a simple, non-tracking collector that does not follow the sun position of the early 90s of approximately 50%, results in a yearly gross heat yield of at least 600 kWh per m² collector area.

A collector surface area of just 6 m² is sufficient to supply warm water to a one-family home with five people using 250 liters of water at a temperature of 55°C. In order to be able to bridge periods of poor weather, an appropriate insulated storage should hold approximately 450 l of warm water. The solar cover value of such a system amounts to approximately 55%. This means that for more than half the time, no additional heating system is needed. These calculations can be recreated by the students themselves if given the characteristic figures and tables.

If heating in a larger collector field is included as a second step, this will save up to 50% of non-renewable heating energy [oil, gas, coal], or half of the valuable fuel wood (comp. Fig. 8.11).

Considering again a one-family home of five inhabitants, a 20 m² installed collector area is sufficient to achieve a solar coverage of 30% using a storage of 1700 l and a heating supply water temperature of 45°C, based on a total yearly energy requirement of 22 megawatt hours [MWh] for both warm water and heating. In these calculations, it is crucial that the solar coverage of warm water production in summer rises to more than 70% and almost no additional heating is needed during transitional periods [October/November and March/April]. This is most attractive in combination with a central heating system based on wood, which is designed for a combustion time of several hours. This example may lead to the realization that a clever combination of alternative facilities may lead to high efficiency.

However, as convincing as this concept may sound, it is necessary to express and examine a series of objections. It is often stated that the construction of such units is

process water heating

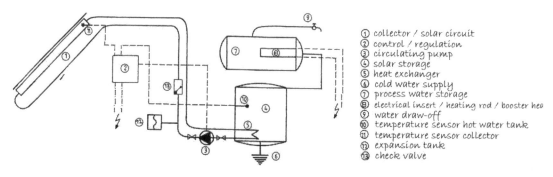

① collector / solar circuit
② control / regulation
③ circulating pump
④ solar storage
⑤ heat exchanger
⑥ cold water supply
⑦ process water storage
⑧ electrical insert / heating rod / booster hea
⑨ water draw-off
⑩ temperature sensor hot water tank
⑪ temperature sensor collector
⑫ expansion tank
⑬ check valve

*Fig. 8.10:* Schematic representation of a solar water heating system with a booster heating for periods of poor weather and winter. (Schmutz, H. 1987 and Schweiz. Energiestiftung SES. SOFAS)

unprofitable or economically not justifiable. As there is a lot of manual labor involved, this is naturally associated with high investments, especially in the case of remodeling.

A solar system costs around $1,350, [1999] per m² installed collector area. A calculation of the prices for heat conversion in 1993 in the US was $0.03 to $0.07 per kilowatt hour for heat from oil or gas combustion. Electric heating cost between $0.07–0.18. A solar device for warm water and heating, such as the one discussed above, produces one kilowatt hour at about 0.3 to 0.5 Swiss Francs, or put into different terms, financial excess costs of approximately 10 Swiss Francs per month. Excess costs were calculated in such a way that the yields from solar energy were subtracted from the interest and amortization costs of the invested sum. This case becomes interesting when we recall

that solar energy is the only energy form that already carries all of its expenses within the installation costs. This is the case because conventional energy conversion causes indirect costs, which are not yet economically assessed and are therefore not included in the sales price. These are the negative impacts on the environment: pollution, heating of the atmosphere, respiratory diseases and accidents during transportation. Students catch on quickly that additional expenditures in craft labor constitute an advantage in terms of the job market—and might be more interesting and economically more successful in the overall picture.

Something emerges in conversations with owners of solar devices that especially catches the interest of 11th graders. The owners speak about the resulting quality of living. They are, for example, very happy

## process warm water preparation including heating

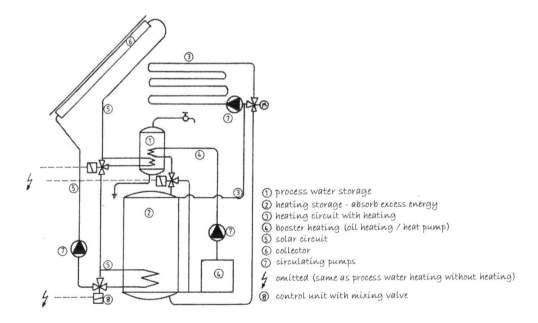

① process water storage
② heating storage - absorb excess energy
③ heating circuit with heating
④ booster heating (oil heating / heat pump)
⑤ solar circuit
⑥ collector
⑦ circulating pumps
⚡ omitted (same as process water heating without heating)
⑧ control unit with mixing valve

*Fig. 8.11:* Schematic representation of a solar system processing warm water and supporting the heating system. (from Schmutz, H. 1987 and Schweiz. Energiestiftung SES. SOFAS)

when they can wastefully take a long shower with the "free gift" of warm water after a warm summer day. They actively experience the coziness of a warm living room because of their previous investment of thought and determination and powers of persuasion into the device. The "interest rate" of this "asset" feels positive as it generates consciousness and a zest for life. Another element is a certain level of independence: The owners are at least partially able to disconnect from the oil or electricity business. This is the topic of a proper decentralization.

A second objection to solar heating is that much more energy is contained inside of the alternative devices than what can be converted [embodied energy]. However, it is just the other way around, as already shown in "Embodied Energy and Harvest Factor," p. 120. Modern wood combustion has the highest harvest factor of more than 7; a solar warm water processing plant shows the second highest value factor of 4. All conventional, so-called economically reasonable and cheap devices show consistently smaller harvest factors of less than 4. This means that the total embodied energy is higher than all energy conversions during the lifetime of the device.

Students often ask the legitimate question: Why are there so many conventional heating systems? The answer is in chapter 8 about crude oil extraction: Oil prices are much too low. A hundred years ago, one barrel crude oil cost approximately $10; it rose to $40 during the Middle East wars and is today at $18. Due to inflation over the last sixty years, the value of the US Dollar fell several times over, which means that crude oil actually became cheaper in comparison to most other consumer goods. Real prices for heating oil fell by 40% between 1976 and 1996.

A third oft-heard argument is that the installation of solar devices is left to a small elite who own their own homes. This argument can be countered when the students become involved in calculations of how much crude oil and nuclear energy could be saved by a consequent use of solar collectors in apartment houses.

This is an important issue in Switzerland as the total population of approximately 7 million Swiss live in 3.3 million apartments. This means that on average, there are about 2.5 apartments under each roof, with an average roof area of 45 m² thus available per 100 m² apartment size. From these, about a quarter, or 11 m², is suitable for the installation of collectors. 36 million m² of installed solar collectors with a moderate calculated gross heat gain of 500 kWh per m² can produce a potential of 18,000 million kWh. All Swiss households together need approximately 49,000 million kWh. From this number, 96.5% are used for energy conversion [warm water and heating], which consequently results in 47,300 million kWh. The calculated potential for solar energy of 18,000 million kWh corresponds to 38% of the above amount. The above-mentioned total of 49,000 million kWh usable energy is presently gained from 75% non-renewable energy [oil, gas, coal], 20% electricity and 5% district heating and wood, respectively. With a consequent use of solar technology, the present proportion of non-renewable and carbon dioxide producing energy could be lowered from 75 to 47%.

Corresponding calculations in the area of industry, agriculture and services would show potentially comparable savings. Project "Solar 91" that was elaborated by the most important Swiss solar energy institutions states similar results and projects that this scenario may be realized by 2025 [Solar '91; guides for the US are available from the US Department of Energy at http://www.energysavers.gov/information_resources/; Solar Radiation Data Manual for Flat-Plate and Concentrating Collectors: http://rredc.nrel.gov/solar/pubs/redbook/—tr.]

## Converting Sunlight into Electricity

Another option for energy conversion that is well suited for discussing with the students is photovoltaic, the conversion of sunlight into electricity using *solar cells* or solar panels. This technology was first used in the US in 1954. Sunlight that hits a semiconductor material such as a specially-coated silicon, generates an electric potential between the lower and upper surface of the semiconductor plate. The direct current produced by the sunlight may be stored in a battery or can be converted into alternating current using a static converter and can thus be fed into the public grid. If a line connection is available, the grid is better storage than a battery. If the system converts more light into electricity than it uses, the current is sold to the power company using an electricity supply meter. When the solar cells cannot not work sufficiently during poor weather and night times the make-up current is taken from the grid and paid for.

Again the question about productivity arises. With ordinary solar panels [12% efficiency], approximately 10 m$^2$ are needed to reach a power of 1 kW in electricity. This corresponds to an investment cost of $10,000 with even higher efficiency. After subtracting losses, approximately 1200 kWh electricity could be produced per year. If calculating with a low device life expectancy of 20 years, 1 kWh electricity costs approximately 0.8 Swiss Francs [1995; $0.4 in the newer US example]. Compared to the 1998 peak rate of 0.21 SFr. per kWh from the grid, this is about four times more expensive [peak rate in Michigan approximately $0.15, or 2.5 times as expensive]. Therefore at first glance, the system is highly unprofitable. Another important part in these calculations is whether the excess electricity fed into the grid can be sold at a good rate [peak rate], which is the case in Switzerland but not necessarily with all US power companies. Plus the calculations do not include interest on the invested assets.

Again it is important in our considerations to realize that the installation of solar cells on pre-existing rooftops or walls does not impact the environment at all, and that there is no need for renewable or non-renewable resources during their operations. Such decentralized systems are autonomous and the longer their life expectancy, the cheaper they can produce.

The claim that more embodied energy is contained within the production of solar panels than the entire energy conversion potential over their life cycle has been definitely disproved by the investigations of the Swiss National Funds in 1996 (May 1996). The production of the best selling polycrystalline cells uses as much "embodied energy" as the cells will produce within 1–2 years [according to other sources: 3–5 years]. The overall harvest factor (refer to chapter 8) of a photovoltaic system is surprisingly high at 6.2, surpassed only by solar warm water preparation plants gaining a harvest factor of 11.2 [calculated with Swiss electricity consisting of 40% nuclear energy, 25% run-of-river power plants and 35% storage power stations].

Taking into account that the prototypes of modern solar cells enter mass production with an efficiency of up to 25%, the price for solar electricity falls from 0.8 to 0.4

SFr., which is almost the rate for peak time energy produced by hydroelectric power. An invention of the Australian photovoltaic researcher Martin Green, plunging the metallic fibers of the power lines and using diffuse solar radiation with the help of an underlying plastic sheeting, could lower the price for mass production of cheap cells containing impure silicon still reaching an efficiency of 22% to about a twentieth of today's cells [newer developments: thin film and nanoparticle solar cells reaching prices as low as US$ 1 per watt: Abound Solar http://www.abound.com; Innovalight: http://www.innovalight.com—tr.].

### How Much Land Area Already in Use Is Available for the Installation of Solar Cells?

This question can also be used to conduct interesting calculations with the students. Data and information used in the following originate from the manual of the "Solar 91" study group, which worked on solar projects sponsored by the Swiss Union for Solar Energy [Schweizerische Vereinigung für Sonnenergie], the Swiss Greina Foundation [Schweizerische Greina-Stiftung] and the Tour de Sol [Solar '91]. Alongside the highway Felsberg-Chur a highway pilot project of a photovoltaic system has been developed, which was known for a long time as the largest solar device of this kind in Europe. The highway runs along the valley floor in approximately east-west direction. Solar panels are installed on existing noise barriers on a length of 700 m. Its peak performance is 100 kW. Approximately 600 km of Switzerland's total of 1700 km highway run in east-west direction and could thus be used to support solar panels. On about 200 km, double noise barriers could be calculated. 800,000 m$^2$ of solar cells with a power of 120 kWh per m$^2$ produce a potential of 96 million kWh per year. If, in addition, half of all parking lots were covered with solar cells, a measure that would cause no additional land wear, this potential could be raised by 720 million to a total of 815 million kWh per year, which corresponds to at least 1.5% of the total electricity production of Switzerland or to the yearly current demands of 150,000 households.

To supply a one-family home with solar current, a cell area of 45 m$^2$ is necessary. In combination with a heating supporting warm water production plant of 20 m$^2$ collector area, a south-exposed rooftop of 65 m$^2$ is needed, which is often provided in conventional one-family homes with a garage building.

How much rooftop is altogether available? It is estimated that Swiss households add up to 150 million m$^2$. Other types of land use in settlement areas add another 2700 million m$^2$. You see that: Installation of 45 m$^2$ solar panels per household would amount to only 5% of the total settlement area. Complete photovoltaic electricity supply to all Swiss households would not require a single square meter of cultivated land or ground.

What would be the price for this remodeling? The necessary 135 million m$^2$ of devices would cost approximately 270 billion SFr. [1991], with a realistic chance to take advantage of a savings due to mass production, leaving a realistic total of 100 billion SFr. Within the next 30 years, the Swiss electricity industry plans to spend at least 40 billion SFr. [new building of five nuclear power plants and rebuilding of hydroelectric plants]. Consumer prices for a total of 30 x 16,300 million kWh amount to 70 billion SFr. It is easy to calculate that photovoltaic systems operate cheaper if calculated with a operating lifetime of 30 years and a low interest rate on the invested money.

### Wasted Energy or Energy Saving

Wasted energy is on the one side unused, released energy such as the heat from exhaust, waste gas, drain water and brakes. On the other side, it is the unused, annihilated exergy of energy, such as steam power lost through a reducing valve. Exergy is the product between energy and its value. Therefore, the value of an energy should not be reduced in conversions without good reasons (comp. cogeneration, p. 138 and Fig. 8.14). It is better to simplify this matter for the students by discussing the insulation of buildings and cogeneration of heat and power.

### Insulation of Houses: Low Energy Houses

The statistics of energy use show that up to 50% of energy conversions are used for heating. (Schweiz. Gesamtenergiestatistik, Swiss Statistical Summary) For a long time little attention was paid to the insulation of buildings because oil prices were low and the "ghost" of climate change was not yet in sight.

First of all, low energy or zero-energy houses built in the often foggy Swiss Midlands during the last couple of years show excellent insulation characteristics. Triple-glazed windows reduce the heat demand by 30%. An additional savings of 20% is gained by excellent thermal insulation of walls and roofs. In homes with additional heat recovery devices, the heating demand is reduced to 35%, compared to conventional new buildings, which conform to the new thermal insulation guidelines of the building code. The heat recovery devices take the heat from drain water and exhaust of a house, which were added to them inside of the house, and recirculates it into the heat circuit.

### Passive Solar Energy Use: Winter Gardens, Trombe Walls

For designing low energy houses, an architect will certainly use the greenhouse principle, that means incorporating as many south-facing windows as possible, in order to receive sun during winter and shade in summer or allow for being received in summer. A minimum of windows will be located on the north-facing side (see Fig. 8.12).

An increased greenhouse function is provided by a winter garden, with glass screens that permit sunlight into the house but do not allow the heat to radiate back out, often enlarged by adding a glass porch to the house front. A ventilation system takes care that in winter the generated heat is moved into the house and in summer directly exhausted to the outside.

Until recently little used, a *Trombe wall* is a sophisticated option of heat storage using an outside wall. Air that was heated up between a glass pane and a storage wall behind it rises and enters the building through an open valve. This results in circulation when the cold air on the floor of the room can enter the interspace between pane and wall from below through another valve (Fig. 8.13). When the sun is not shining, circulation ceases and valves are closed. During the night, the heated house wall slowly emits its heat into the house.

After describing the winter operation of a Trombe wall, we will need to show its cooling function in summer. Another air valve at the cool north side of the house or apartment allows cool air to flow into the room. The air escapes from the lower valve of the Trombe wall into the area between wall and glass, is heated up and released to the outside through a opened valve at the upper end of the glass wall. During this time, the upper valve of the storage wall is closed. This results in a cooling circulation, which can be enhanced by a small fan.

### Combined Heat and Power [Cogeneration]

A lumberjack works in the forest during a cold winter. Physiological combustion processes convert energy stored in glucose into kinetic energy. This process also releases heat, which the lumberjack appreciates very much. If this principle were applied in a decentralized way, we can fall back on the technically sophisticated cogeneration machine.

To simplify things, we will describe its operation in a mountain hotel that is not connected to the grid. Data were obtained from the documentation *Energy-saving Building Systems* in Switzerland, 1980. Due to the lack of a natural gas pipeline, electricity is produced by a diesel generator with a power of 142 kW. Of 100% energy of the fuel [diesel] 32% is converted into electrical energy, and the remainder is emitted unused into the environment in the form of radiation [10%], exhaust heat [36%] and cooling water heat [22%]. This electricity conversion machine has an efficiency of 32%, which is comparable to the efficiency of a modern nuclear power plant. Heating is produced by conventional oil heating with a boiler size of 150 kW, which converts 100% heating oil energy at a maximum of 70% into usable heat; losses of 30% are emitted into the environment as radiation and warm exhaust (comp. Fig. 8.14).

If the hotel is converted to a total-energy plant with a power of 142 kW [corresponds to a middle-class car engine], a completely different energy balance emerges. The generator driven by a diesel engine converts 100% fuel energy into 32% electric energy. A high percentage of the heat from the cooling water [22%], waste gas [36%] and radiation [10%] is regained and can be delivered to the heating and warm process water via a heat exchanger. The loss of 15% is due to the exhaust gas that cannot be completely cooled down. The efficiency of the plant increases to 85%, the capacity of the useful energy gain increases from 150 to 236 kW, providing much more warm water without any additional costs. Heating oil is no longer needed at all.

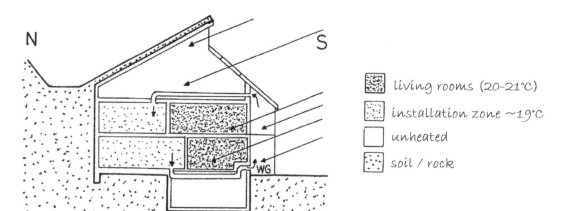

*Fig. 8.12:* Schematic representation of a low energy house with a winter garden [WG]. Besides excellent insulation, most emphasis is put on the passive use of sun energy [south alignment, large glass areas]. The soil-rock banks are used as long-term storage. In addition, it is easy to include a Trombe wall (comp. Fig. 8.13, from Schmutz, H. 1987).

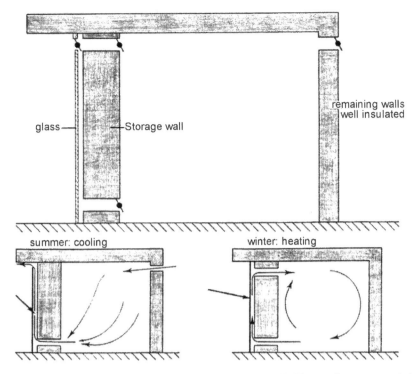

*Fig. 8.13:* Function of a Trombe wall shown in a section drawing. Controlled by simple opening and closing of air valves, the wall can effect cooling in summer and heating in winter. (from Infosolar 1991)

This example shows that high-grade energy such as crude oil can be used multiple times during conversion without wasting the high value [thermal discharge]. The application of combined heat and power plants is especially suitable in the case of smaller buildings and industrial plants. Larger-scaled plants can address noise and exhaust problems better by using natural gas engines (compare to Wärme-Kraft-Koppelung in dezentralen Anlagen [Combined Heat and Power in Decentralized Plants] 1983; see also publications of the Lawrence Berkeley National Lab, Electricity Markets and Policy at http://eetd.lbl.gov/EA/EMP/emp-pubs.html).

## A Look at the Global Energy Problem

### Energy Justice

In most of the previous reflections, Switzerland was used as an example for various energy scenarios. The German energy problems are summarized in *Competent for the Future?* (Loske, Reinhard, and Bund für Umwelt und Naturschutz Deutschland, 1996). In the following we want to look at the entire world. How does the way Switzerland deals with energy relate to other countries? In terms of climate, Switzerland is average in the number of annual heating days. An average Swiss converts 40 times more energy per year than he does physiologically in his own body energy. In other words, he has 40 energy slaves. In 1993, this was equivalent to 3700 kg oil units. Comparatively, a US citizen uses 7760 kg or 78 energy slaves, and an Indian

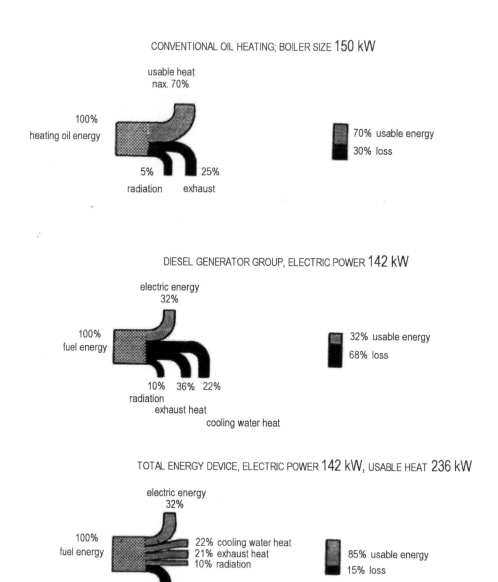

CONVENTIONAL OIL HEATING; BOILER SIZE 150 kW

usable heat
nax. 70%

100%
heating oil energy

70% usable energy
30% loss

5%    25%
radiation  exhaust

DIESEL GENERATOR GROUP, ELECTRIC POWER 142 kW

electric energy
32%

100%
fuel energy

32% usable energy
68% loss

10%  36%  22%
radiation
exhaust heat
cooling water heat

TOTAL ENERGY DEVICE, ELECTRIC POWER 142 kW, USABLE HEAT 236 kW

electric energy
32%

100%
fuel energy

22% cooling water heat
21% exhaust heat
10% radiation

85% usable energy
15% loss

53% heat recovery

15%

exhaust heat
exhaust gas (unusable part)

*Fig. 8.14:* Comparison of a total energy plant and a combination of oil heating and diesel generator. Combined heat and power [cogeneration] reaches a much higher efficiency ( from Energiesparende Gebäudesysteme in der Schweiz [Energy-saving Building Systems in Switzerland], 1980)

citizen uses 235 kg or 2 energy slaves. A Nepalese citizen living in the cold highlands is entitled to only 20 kg oil per year—that is 185 times less than a Swiss! These facts lead to the consideration of whether provisions should be made within a reasonable time that all citizens of the earth may reach an energy consumption and a standard of living similar to the Swiss. Corresponding calculations lead to the result that this would triple the global energy expenditure—in sharp contrast to the current efforts to reduce the yearly $CO_2$ emissions. This is so because more than 90% of the current energy conversions is based on $CO_2$-producing non-renewable fossil energy carriers such as oil, gas and coal. It might be possible to deliver the excess energy using additional nuclear power plants, as suggested by the large energy providers. Another calculation run by students will show that this would require an increase of nuclear power plants globally from 440 today to 16,000 plants, which would be an investment of approximately 80,000 billion SFr. The consequences of this alone in terms of uranium mining are briefly described on pages 128–129 and in Fig. 8.7.

This imagination of a dead end in using today's centralistic large scale plants becomes more and more obvious. The way leading to a practicable energy justice is based on three pillars: 1) reduction of the waste of energy in industrialized countries; 2) development of decentralized alternative energy technologies; and 3) ideological and financial support of the utilization of alternative technologies in Third World countries. Testing gentle alternative technologies in Switzerland, using renewable energy carriers that produce the least amount of $CO_2$ possible, is therefore not necessarily an act of selfishness but may lead to a real improvement of living conditions worldwide.

## Technocratic Planning an Illusion?

Switzerland may serve once again as an example of technocratic energy planning. In the context of the controversial issue of building new nuclear power plants, the Swiss All Energy Commission was in charge to calculate the energy demand until the year 2000. The majority of members of this committee, appointed by the Federal Council [Bundesrat], were either representatives of various branches of the power-supply industry or scientists, who cooperated intensively with the energy industry. Based on the calculated increase from 290 petajoule [PJ] total energy demand [final energy] in 1960 to 960 PJ in 1975, the 2000 demand was predicted to be 1510 PJ. This number is already a clear reduction compared to the 1972 estimate of 2050 PJ by the Confederate Office for the Energy Economy.

This was not a surprise as the energy crises in connection with the war in the Middle East in 1973 limited the waste of energy for the first time (Fig. 8.15). Due to the interventions of environmental groups that became politically more active, the Swiss All Energy Commission had to also calculate scenarios based on energy saving and energy waiving. In 1978 they calculated a saving scenario of 1130 PJ. The environmental groups presented also an energy scenario for the year 2000, which was calculated without changes in the quality of living but with a more economical and efficient use of energy including alternative technologies, stating that 702 PJ may be sufficient. As early as 1975, opponents of nuclear power presented a project that estimated 880 PJ in the year of 2000.

What actually happened? After energy conversions rose to 830 PJ in 1991, they reached 786 PJ in 1994, clearly marking a reduction. They rose again to 807 PJ in

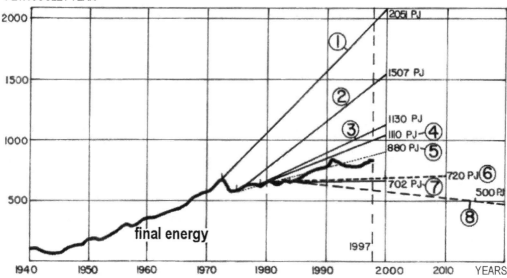

*Fig. 8.15:* Scenarios of the final energy consumption in Switzerland are disproved by actuals. The bold curve shows the effective final energy conversion in Switzerland between 1940 and 1997 in petajoule [PJ]. Thin lines illustrate the following scenarios: 1) Federal office for energy [Bundesamt für Energiewirtschaft] 1972, 2) All Energy Commission 1975, 3) Final report of the All Energy Commission 1978, 4) Confederate Energy Commission [Eidgenössische Energiekommission] 1979, 5) Scenario by the opponents of nuclear energy [AkW-Gegner EWU] 1975, 6) Confederate institute for local, regional and national planning [Eidgenössisches Institut für Orts-, Regional- und Landesplanung] 1984, 7) Project "afar from practical necessity" of environmental organizations [Projekt "Jenseits der Sachzwänge" der Umweltorganisationen] 1978, 8) Engineering firm INFRAS assigned by the WWF [Ingenieurbüro INFRAS im Auftrag des WWF] 1984; saving-scenario including abandoning nuclear power plant technology until 2020 [Sparszenario mit Ausstieg aus der Atomkraftwerk-Technologie bis 2020]. (all according to Strahm 1992; TA 7/25/1986)

1996, which was mostly due to many heating days during that year. It was most of all the industrial sector that was able to clearly reduce their energy needs and use energy more efficiently, including new technologies. If the trend to a more responsible energy use continues into 2000, a final energy expenditure not exceeding 800 PJ is estimated, which is only slightly more than half of the number calculated by the GEK in 1975. [Switzerland consumed 859 PJ in 2000, and 865 PJ in 2007—tr.]

This result may be viewed as an encouraging sign that the tireless promotion of gentle energy technologies and energy-saving proposals has gradually fallen on fertile ground. Switzerland is seen as a pioneer in the development of solar energy and the building of zero-energy houses. However, most sustainable proposals and implementations originate from mavericks and small enterprises, not big business.

## Pedagogic Considerations

Analyzing the course of energy economy blocks between 1987 (the year of the alarming accident at the Chernobyl nuclear power plant) and 1997, one problem becomes clear: Young people today are taking the current situation very seriously however large their differences in knowledge about the situation. It is therefore necessary to discuss the drama of the current ecological and political relations to wake up the students who are still dreaming. On the other side, it is a balancing act, because already the more awake students may develop anger, resignation or depression, if they are exposed again to what they already know. Pointing out what we adults did wrong has a paralyzing effect because the young people take these issues very seriously. It is liberating when we then discuss the possibilities of turning around to a forward-looking energy policy. It is very important that the alternatives introduced are not perceived as mere symptomatic or as privileged botched-up job.

The path of energy saving technology, solar energy as well as sustainable wood use becomes a potential reality for the students due to the evidence presented. It is not only for the privileged to walk this path. The only step necessary is to change one's ideas, and it can be fruitful as long there is at least some money left. Another important pedagogic aspect of discussing alternative technologies is that these technical devices are relatively easy to understand and thus allow the application of elementary physical phenomena in a beautiful way. Examples of successful student quarterly and graduation projects show the possibilities of a practical implementation of the ideas in everyday life (comp. Fig. 8.16), and the demonstration of a warm water processing device as a final project is much more meaningful than any number of technical descriptions.

When concluding, I always stress the importance of alternative devices being decentralized. This way they can become reality from the bottom up, initiated by individual decisions. The modern principle of *ethical individualism* can be demonstrated and experienced in an example. When we are talking about centralized plants such as nuclear power plants, implementation assumes many preliminary bureaucratic and political decisions that a citizen may not be involved in; he/she has to delegate them to the technocracy as an instrument of action. The student has the chance to experience, how energetic actions can be done where an individual lives and consumes. This kind of "doing" is always combined with "learning" and the spreading of the knowledge gained and success stories. This is possible because the alternative technology is understandable.

Another aspect of dealing with solar power is its active context with the environment, which leads to reverence and respect of the world. If it is possible to guide the class to such experiences, this can lead to what teachers call an "hour of glory." And the student may write a short commentary into his report book, expressing that he experienced his schooling as timely and relevant to the real world, and that important questions were finally seriously discussed.

The economy of energy is not only a key subject in the representation of the entire global economy, which shall be at the core of the 12th grade block (refer to chapter 10). It is also an understandable example of the development of a way of living that is in keeping with the conditions of nature and helps to cure existing injustice. This is because the sun as an energy source is available for everybody.

It is likewise possible to put the inventive genius at everybody's disposal. There is no need for exploitation—of peoples or nature. In this way, the students experience how an energy policy which is in harmony with the processes of nature and the ethical basic needs of human society is at the same time a *policy of peace*.

# Process warm water preparation:

## Schematic representation of my solar device

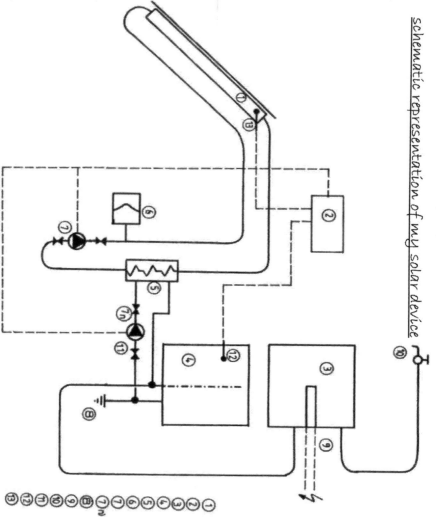

Fig. 8.16: Schematic representation of a solar device that was constructed in 1987 as a graduation project of a 12th grader. It works flawlessly [1997]. (Schmutz, H. 1987)

① collector / solar circuit
② control / regulation
③ process water boiler / storage
④ solar boiler / storage
⑤ external spiral heat exchanger
⑥ expansion tank
⑦ solar circuit circulating pump
⑦ₙ process water circulating pump
⑧ cold water supply
⑨ electrical insert / heating rod / thermostat
⑩ water draw-off
⑪ various sanitary fittings
⑫ temperature sensor boiler
⑬ temperature sensor collector

# Chapter 9

# Paleontology/Anthropology—Twelfth Grade

**About Human Science [Anthropology] in Grade 12 and the Selection of Topics**

Entering grade 12, adolescents have already passed through the most important preliminary exercises leading them to an independent formation of their judgment. In grade 9 they exercised how to move from an observation of tangible objects to an appropriate judgment within the realm of the inorganic world. They learned to think in chains of cause and effect, guided by sensual observations. In grade 10, students learned to distance themselves from the tangible objects and study processes of the dynamic world, which requires a more autonomous and more flexible thinking. Differences in space and time can be imagined; movement as such can only be thought. In grade 11 they focused on thinking about the living world as being full of soul. The manifold aspects of the concept of intentionality was central to all exercises (comp. chapter 1).

In this way we teachers could first explain the world that has become, and then the developing world, which includes all life processes, could be included into the guided formation of judgment. Finally we asked about the impulses and directions of the constantly changing life. Now in grade 12, the pressing questions are about the motives of development, to get to the bottom of the sense of life and think it through. The material that is now brought over to the students serves predominantly to work on the free creation in such a way that allows the development of realistic judgments. If Steiner demands that grade 12 should provide an overview of as many world areas as possible, this also means that the future can be viewed only from an overview of what has become up until now.

As there are no directly practicable indications from Steiner, nor an esablished syllabus (refer to Rohrbach 2000), we will explain the selection of the following materials and topics. Within the conferences of the first Waldorf school in Stuttgart, Steiner proposed that we could discuss the *movement of the continents*, to make clear to the students how large islands and continents moved on the spherical surface of the earth during geological time scales and how they are controlled by fixed stars. (*Conference with Teachers* on 04/25/1923; Steiner 1924b) When Steiner made this remark in 1923, the scientific world did not have any idea about plate tectonics. The only existing publication about this topic, *The Origin of Continents and Oceans* by Wegener, 1919, stating that the continents would move since the Mesozoic, was judged by contemporary experts as absurd. It is interesting that Steiner alluded to theis paper despite the fact that Wegener was not discussed at that time and his ideas were not revisited until 1982 during a large Alfred-Wegener conference. In addition, Steiner suggested to study ethnology [cultural anthropology].

What do both indications have in common? A review of the history curriculum in grade 10 may be helpful. It is the task of the history teacher to show the students how entirely opposite landscapes such as Mesopotamia and the Nile valley may have beneficial or hindering conditions to specific cultural developments. The becoming becomes understandable through what had become. If change in the earth's history is the topic in grade 12, it would be worthwhile to ask questions about the beneficial and hindering conditions of the paleocontinental living areas on the evolution of plants, animals and the human world. In this way Steiner's suggestion about the changes in the position of continents and oceans makes sense. It would be appropriate to present and develop global and regional paleoecology. We do not have to search for the driving forces of changes in the earth's "face" in the darkness of coincidence but in cosmic impulses that reveal themselves in geometric order principles that we already encountered initially in the geology main lesson in grade 9 (tetrahedron structure, chapter 2) and in astronomy in grade 11 (chapter 7) but could not thoroughly penetrate them in thinking. The presented main lesson design wants to risk first steps on this outlined path to approach the goal of Steiner's curriculum indications—although we are still far away from solving the cosmic-earthly relations of evolution. (Steiner 1921)

One serious objection should be broached at this point. It is right to say that the adolescent should get acquainted with the modern cultural world, that all instruction in high school should be *life skills*. From this premise it is often deduced that *economy* should be a topic in both 11th and 12th grade geography (refer to the contributions by Kübler 2000b, and Göpfert 1999). This argument receives support also from the fact that public education teaches economy in the upper grades, rendering it suitable for the high school diploma in Germany and Switzerland.

Certainly the adolescent has the right to learn about modern economy and civilization. However, in discussing this controversial topic, the sense question is very much in the foreground. Modern economic life is joyfully exploited by some and felt to be unfair and unjust by others, leading to the desire for a fair social and economical structure for both human beings and nature. However, if one agrees that the 12th grader should for the most part learn humanly appropriate judgments, the teacher has to look deeper at the basic principles of his/her instruction. These are only then sufficiently given when the *past* processes of the natural- and cultural evolution have been sufficiently appreciated and thoroughly penetrated by thinking. Decisions for the future need the appreciation of the past and extraction of insights into the nature of the past and current development. In this perspective, the paleontology/anthropology main lesson is an essential precursor to be able to give the social and economical present and future serious thoughts for responsible judgments and actions. I am, therefore, not disapproving an economy main lesson in grade 12 but highly suggest a precursor block (chapter 10), and the main lesson presented here should be the major geography main lesson in grade 12.

In a global ecological view we must take a stand about mineral resources and fossil energy sources. This can only happen in a responsible way if the formation of these resources throughout geological time has been discussed before. This was begun during the energy main lesson in grade 11 (comp. chapter 8). If modern man intervenes and reshapes the fabric of the world that has become, he can do this in a balanced and appropriate way only if he/she has intensively studied the nature of the earth and its prior development. It is interesting to see that especially many geologists in the last decades have admonished about premature technologies that have changed the world and helped to correct several wrong decisions (e.g., the activities of Alexander Tollmann against the nuclear power plant in Zwentendorf, Austria—the first-ever national referendum and atomic energy prohibition act worldwide).

If we demand such a way of judgment and decision education, we as teachers have to set our students good examples in creating syllabi and as human beings. Viewed in such a perspective, the design of the curriculum becomes a highly moral and responsible task.

## About the Characteristic Style of the Twelfth Grade Block

If the task is to provide an overview of an entire area of the world, you will preferably pay attention to two aspects. On the one side, all that was done in earlier classes should be reviewed in a skilled way. Main lesson books written by the students in their own words are an excellent tool for this, as students connect with their own earlier performance and experiences during re-reading. On the other side, it is most important to highlight the many and diverse relations during the presentation of the material. This is where the importance of geography classes becomes most obvious. It is the nature of geography to connect all three realms of nature with each

other and with the realm of human culture. We could say that, especially in grade 12, the focus of all natural and cultural science classes should be geography. The material following focuses on paleontology and anthropology as possible examples of this principle and comes from the design of this block over many years in the Wetzikon school. A detailed presentation of the 12th grade block is now available from the education research group, department, Kassel. (Schmutz et al. 2004)

## A Preliminary Remark about Geological Dating

Geology uses a time scale that spans over four billion years. Periods during which organisms become evident and changes in the earth's shape are embedded into this time scale that uses today's solar year as reference unit. Creationists and also some anthroposophic circles disagree with this form of dealing with millions of years (compare to Bockemühl 1999a; Schneider 1982). The following presentation intends to provide the reader with the opportunity to come to a differentiated examination of time, resulting in a subsequent ability to use the geological time scale. Geology uses various methods of dating and continuously works on correlating working method and results. This leads to more reliable specifications.

### The Method of Determining Seasons

If you study lime sludge deposits in stagnant waters, such as a postglacial lake at the margins of the Alps, over the time course of a solar year, you find a brighter following a darker sludge layer. The brighter lime is formed during the season which degrades dead plankton of microorganisms with calcareous shell as completely as possible. The remaining pure white lime sludge is deposited at the bottom of the lake. If from autumn on, more plankton dies than can be degraded by microorganisms, the deposited lime sludge still contains organic, carbonaceous remains that are darker colored. Within a core-drilling of 100 m in the alluvial deposits of Lake Zurich, more than 10,000 years can be counted despite substantial changes in the thickness of the layers compared between years. (Kelts 1978)

The method of *dendrochronology* is very similar. Tree growth rings with their characteristic light-dark patterns present also an image of the seasons. Additionally, the thickness of tree rings reflects the approximately 11-years rhythm of sunspot activity. This is even further supplemented by climate fluctuations of longer periods. Comparative studies of the annual growth rings from live and dead trees has resulted in a combined scale of approximately 17,000 years. (Wagner 1998)

Arctic and Antarctic ice cores provide another further back-reaching time dimension. Measurements of the isotope ratio between $^{16}O/^{18}O$ in gas occlusions of ice masses, which were deposited above each other year after year without human interference, document the extreme seasonal changes at the polar regions between a warm, light summer to a cold, dark winter. We are now able to archive complete ice cores of more than 3 km length, which allows a count of more than 270,000 years, based on a year as described above. Seventy years after Steiner's emphasis on the importance of the Great/Platonic Year for the understanding of evolution, we can read more than ten of such Platonic Years of each approx. 25,920 solar years within these earthly documents. During this time, no erratic or serious changes in glaciations were found besides changes in global temperature and global carbon dioxide content measured as changes in gas composition. More details about this method can be found in several short reports. (Steffensen et al. 2008; Stauffer and Stocker 1995; Stauffer 1993; Schwander et al. 1993; Jouzel et al. 2007; Blunier et al. 1998)

### Stratigraphic Dating

In several locations on earth—such as the Colorado plateau, opened up by the Grand Canyon—a sequence of alternating marine and continental deposits remained undisturbed in a depth of several kilometers. This is not only the case for the layers of the Cenozoic, Mesozoic and Paleozoic, but in parts also for the Precambrian. There are gaps between layers due to erosion and degradation. However, by comparing regionally different deposits, a geological overall profile with more than 20 km depth can be assembled! An extended study of the large treasure in fossils of embedded remains of animals and plants shows that very many species lived only during a restricted period of time of 4 MY (= million years) or less. These kinds of fossils that only occurred for a short time in high numbers are called *index fossils*, as they can be used to perform relative dating. Today the stratigraphic time scale includes 13 valid time systems (such as the Triassic), 45 time series from the Cambrian until today (such as Muschelkalk), and 116 time stages (such as lower Muschelkalk = Wellenkalk). Up to now, the long lasting and far back Paleozoic is only very roughly classified. If we find index fossils anywhere in the world that are not associated with a larger set of layers due to motion phenomena, we can assign this probably isolated layer to one of approximately 120 time stages. The time scale developing thus is an expression of the becoming and passing of plant and animal species in the time course of life history of the earth.

It becomes possible to estimate the lifetime of a species by relating it to the mass of today's deposition or disintegration rate of debris or rocks. Observations of young rocks show that for example 3 cm of rocks are eroded per century in the region of the Alps. At a yearly precipitation of 2000 mm, chemical erosion with rain water dissolves approx. 1 cm of pure limestone or 5 cm marly limestone within 100 years. Deposits in river deltas may have a yearly growth rate of more than 1 cm. If we transfer these observational results onto geologically earlier times, we are using the principle of actualism or as it is more commonly phrased in English: uniformitarianism [uniformism]. (Dullo 1999; Reif 2000; Gould 1965, 1987; Baker 1998; Anderson 2007) In principle, today's processes can be compared to those in the past. Actualism is one of the most important methodic assumptions for geological studies. It relates to the constant validity of physical and chemical laws but expects a strong variation in the intensity of life, which affects sedimentation and erosion rates. The book *Earth System History* offers many examples demonstrating the stratigraphic method. (Stanley 2008)

## *The Method of Geochronology*

The foundation for the method of geochronology is atomic physics, which describes radioactive decay chains of unstable isotopes to stable elements. For each decay chain, there is a characteristic time scale called half-life, which is the time that statistically passes until half of an isotope is converted into the next isotope or element. A second assumption is that the isotope ratio returns to the same level after melting or strong metamorphosis. The supporters of this method assume that neither physical facts nor material characteristics have changed during at least the last 4.6 billion years; they strictly assume the principle of actualism, which should go back as far as the first formation of solid rocks (compare to Grant 1999; Bockemühl 1999b).

The measuring procedure involves the crushing and separation of fresh crystalline rocks, resulting in those minerals that contain the isotopes to be measured. The beginning of the isotope decay, that is, the crystallization from magma or cooling of rocks heated by metamorphosis, can be calculated from the ratio between the measured isotope quantities.

Mica-containing rocks are suitable to measure high rock ages due to the ratio between the mother isotope Rubidium 87 and the stable daughter isotope Strontium 87. The crucial half-life in this decay chain over 17 stages amounts to 48.6 billion years.

Most acidic to intermediate crystalline rocks contain the very rare and resistant mineral zircon. With the help of zircon minerals extracted from rocks, scientists can determine the ratio of thorium 232 to lead 208 (half-life 14 billion years), as well as the ratio between uranium 238 to lead 206 (half-life 4.5 billion years) or uranium 235 to lead 207 (half-life 700 million years) respectively. This measuring method was internationally standardized using comparative calibration measurements and is very reliable, especially as crystalline zircon grains can be extracted easily from rocks. If you work with metamorphic rocks and you want to determine the age of the metamorphosis, you can use the isotope decay of potassium 40 to the noble gas argon 40. Again mica minerals are well suited for this. Due to a half-life of 1.3 billion years, and the possibility of argon to diffuse out, this method is suitable for the Mesozoic and Paleozoic.

The radiocarbon method, known and often mentioned by laymen, is frequently used in archeology but is for the most part not suitable for geology due to its inaccuracy and also for the short half-life of the carbon isotope of only 5730 years. Geochronology is characterized by a *clock process*, detectable by statistics and combined with the radioactive decay of isotopes. Its stubborn beat cannot be altered by life processes or human interference. Therefore it is exclusively part of the inorganic world. A critical recognition of this method can be found in the essay "Isotope analyses and age determination in Goethe's 'Primary Rock'—About the Handling of the 'Four Elements' in Geology" (*Isotopen-Analysen und Altersbestimmungen an Goethes "Urgestein"– Vom Umgang mit den "vier Elementen" in der Geologie*, Bockemühl 1999b).

Grant verifies that radioactive decay processes, as assumed by isotope geology, are stable for at least the last 80 million years. This time marker does not constitute an upper limit. In this context, the statements of Wachsmuth and Lehrs should be regarded as not true. (Wachsmuth 1980, 1952, 1950; Lehrs 1958, 1953) Both authors cite part of a lecture given by Steiner in 1905, the written version of which includes patchy notes from the audience, which could easily contain errors. (Grant 1999; Steiner 1905b) Wachsmuth and Lehrs assume that radioactivity has existed for only the last 15,000 years. Bosse puts the beginning of crystallization into the Tertiary, ergo radioactive processes could not begin before 65 MYA [million years ago]. (Bosse 1993) This also is in strong contradiction to the extremely well verified research results.

## Geometric Method of Determining the Sequence of Events after Material Motion

Based on an analysis of the spatial geometry of a geologic opening-up, the sequence of rock-forming and rock-metamorphing events can be determined with certainty. This shall be illustrated with an example (Fig. 9.1). The fault B1 is the youngest event, disrupting the entire stack of layers. The otherwise undisturbed flat deposit A1 is one stage older. Even older is the fault B2, which does not pass through layer A1. Before the formation of fault B2, the granite G intruded from below. On the other hand, layer A2 is older than the granite G and younger than the folded and crest-shaped eroded layer A3. This layer can be further separated into a younger folding event F and an older deposit. In this case, seven events can be determined beyond doubt by logical thinking alone in terms of the sequence in age. It is not possible to think the events in reverse, but there is no information about the time periods between those seven events.

This relative age determination informs us about the biography of the rock complex. An example for the execution of this method

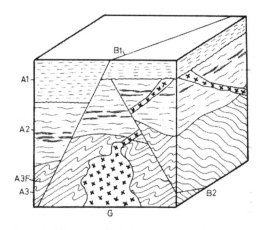

*Fig. 9.1:* Schematic block diagram of a rock complex, visualizing the temporal sequence of seven events. From A1 to G, refer to the text.

is the study, "The macroscopic geometry of the Silvretta nappe pseudotachylites in the NW border of the Lower Engadine window" (*Die makroskopische Geometrie der Pseudotachylite in der Silvretta-Decke am NW-Rand des Unterengadiner Fensters*, Schmutz 1995).

## *About a Characterization of Time*

The measure of the clock registers the course of time in the small picture. Observing the movements of the sun, moon and stars, combined with seasonal life processes on earth, lets us describe time in the large picture as a rhythmic sequence of repetitions. The grasping of metamorphosis in living beings leads to the growth-oriented lifetime. The acknowledgment of the sequences of events and actions in a human life leads to biographical time. These four time qualities are interconnected but each has its own characteristics (Fig. 9.2). I therefore describe these four time-measuring qualities

in the presentation of the four methods of geological dating. When studying the geological literature in terms of statements about time, you can observe that the specifications become more exact the more the four dating methods are related to one another. It is therefore not possible to judge any of these four methods as better or worse, because the meaningfulness and reliability of any dating increases by connecting them with each other. Another amazing result of our work is that, due to the usefulness of geochronology, the principle of actualism is obviously valid until early geological times. In the following we will indicate geological age in million years (MY) according to the level of knowledge in the earth sciences.

The quality of the duration of a year previous or later is not described by this yet. It indicates only that an event with a larger number (in MY) happened before an event with a smaller number.

| Natural Kingdom | Time Quality | | Geology |
|---|---|---|---|
| Mineral | Beat - Time Scale | Mechanic clock<br>*Linear* floating | Geochronology<br>Isotope Geology |
| Plant and Animal | Rhythm - Timescale | Sequence of years, months and days as relations between the cosmos and the earth<br>*Fluctuation* of swelling and shrinking | Dendrochronology<br>Determination of seasons<br>Analysis of sludge deposits and ice cores |
| | Growth - Timescale | *Logarithmic* increasing and leveling of growth and reproductive processes | Stratigraphy<br>Becoming and passing of index fossils |
| Human | Biographical - Timescale | *Consequence* of Events, which create new conditions for new developments | Geometric method to determine the sequence of events |

*Fig. 9.2:* Characteristics of the four qualities of time and their relation to the four different methods of geological dating.

## Exemplary Notes about Paleontology

Provided that the class has already learned about systematic botany and zoology in previous natural science blocks, we can now focus on the sequential steps of organisms and the effect of life processes on the transformation of the earth both geologically and physically. With a sensitivity developed in the astronomy block in grade 11, and with the help of a teacher, students do not only search for a cause-and-effect process but are able to grasp the difficult intentional relationships. These relationships with their mutual interactions between "cause" and "effect" are followed in such a thinking processes that allows the discovery of their inherent teleology [purposefulness, German: *Zielgerichtetheit*].

### *Early Life Processes and First Rock Formations*

On a worldwide hunt for the oldest rock formations, a series of complex metamorphic rocks was found, e.g., in west Greenland and northwest Australia, not as expected and predicted granites or basalts. The radioactive age of the metamorphic transformations of these rocks is around 4000 to 4200 MY, which means that the oldest rocks on earth are younger than the most upland rocks on the moon, which are 4100 to 4500 MY old. (Dreibus-Kapp and Schultz 1999; Wanke et al. 1977; Wanke and Dreibus 1983; Dreibus and Wanke 1990; Taylor 1994; Taylor, Taylor, and Taylor 2006; Taylor 1987; Taylor 2009; Duke et al. 2006; Burns et al. 1990) This may suggest that earth and moon separated before the formation of the first permanent rocks on earth. This result conforms to the indications of research into the spirit by Steiner in *Occult Science: An Outline*, "Man and the Evolution of the World."

The oldest rocks in Greenland show a typical sequence of metamorphic and strongly deformed crystalline rocks. The metamorphic rocks are former sedimentary rocks; the oldest crystalline rocks of approx. 3900 MY are more alkaline, meaning they contain less quartz compared to the slightly younger vulcanites and intruded granites. (Moorbath 1977, 2009) More recent investigations (Hayes 1996, De Gregorio et al. 2009) of old rocks (3800 MY) in the Itsaq Gneiss complex of Greenland have shown that they must be *biogenic sediments* as the measured $^{13}C/^{12}C$ isotope ratio in carbonate fluoride apatite grains points to biogenic carbon.

The oldest rocks in Australia are slightly older (4100 MY; Pflug 1989; Pflug 2001; Pflug 1984) and also suggest metamorphic transformations of older plutonites and vulcanites. Rocks approximately 100 to 200 MY younger form a cyclic sequence which is well known from younger periods. These are typical shallow-water deposits such as sandstone, quarzites, greywackes, clay slate (argillite), carbonate and silica deposits, sulfate rocks and phosphates. Special interest was paid to the sometimes high (up to 3% of the rock mass) carbon content of organic origin. This value is similar to values from current shallow-water deposits.

Based on these and a few other discoveries—for example the Aldan massif in the Precambrian of east Siberia—a hypothesis was put forward in the 1980s that *at the time of the oldest remaining rocks, there was already a flourishing organic life in the oceanic waters.* Up to the present day, this hypothesis has won more and more support. Some aspects of this chain of evidence will be presented below in more detail.

## The Importance of Cyanobacteria

### A Picture of an Early Landscape

Modern geologists construct something like the following picture of the early geological earth: The surface of the earth was mostly covered by shallow oceanic waters. There were no continents like today's. Some dozens of larger and a multitude of smaller islands rose bare and desert-like above the water. There was a lot of volcanic activity—on the islands as well as on the ocean floors. The atmosphere's composition was mostly of various gases released by volcanism: $CO_2$, $H_2O$, $CH_4$, $NH_3$. The atmosphere was probably about 50% carbon dioxide, a few percent water vapor and a lot of nitrogen. There were also traces of hydrogen sulfide ($H_2S$), sulfur dioxide ($SO_2$), hydrogen, methane ($CH_4$), ammonia ($NH_3$) and noble gases. The most striking difference to today's atmosphere was the high proportion of $CO_2$ and the absence of oxygen ($O_2$). In connection with the radiation intensity of the sun, which was approx. two thirds of its present intensity, we can predict a global surface temperature of 100° to 125°C due to the greenhouse effect, and an atmospheric pressure of approx. 9 bar. Due to the high content of water vapor, the atmosphere must have been very foggy and muggy. It is important to imagine that the earth's surface water was mainly in liquid and to a smaller extent in vapor form. This means there was already a separation of land, air and water. For life at that time, the element of water, already slightly flooded with light, was especially meaningful.

Discoveries of a special kind of rock suggest that a multitude of photosynthetic organisms settled the world as early as 3900 MYA. North Pole, a location in northwest Australia is a well researched location of a series of more than 3500 MY

old rocks, exceptionally well preserved and which luckily underwent only minor metamorphosis. (Groves, Dunlop, and Buick 1981) The area around North Pole consists of granite domes of several square kilometers, which are embedded in a changing sequence of older volcanic and sedimentary rocks. In this changing series, arcade-like curved formations of approx. 20 cm height were found in brighter hornstone layers, a fine-grained siliceous rock (Fig. 9.3a). The hornstones are embedded into volcanic basalts showing a pillow lava structure indicating that they were deposited in water. Such formations, the so-called stromatolites (Fig. 9.3e), are known from geologically younger, calcareous rocks. They consist of lime secretions, caused by the photosynthetic activity of colonies of cyanobacteria. If you visit, for example, oxygen-deprived water bodies such as the Dead Sea, or the supersaturated lagoons at the Barrier Reef in east Australia, you can observe extant living cyanobacteria and marvel at the vault-like lime towers of up to 1 m in height.

It is furthermore known that carbonate deposits or biogenic masses such as wood are replaced by silica over the years, so that the old form structure is exactly preserved. A well-known example is the silicified woods from the Triassic. Luckily this process happened a long time ago within the observed rocks, and the carbonate structures transformed into silica remained protected from disintegration until present day.

As there are many bubble cavities, this suggests a shallow-water deposit. The basalts are in part in the interpenetration framework with bright vulcanites of granitic composition and only wind-blown sands, which formed on a solid land surface. From this and a few other pieces of circumstantial evidence, geologists came to a reconstruction

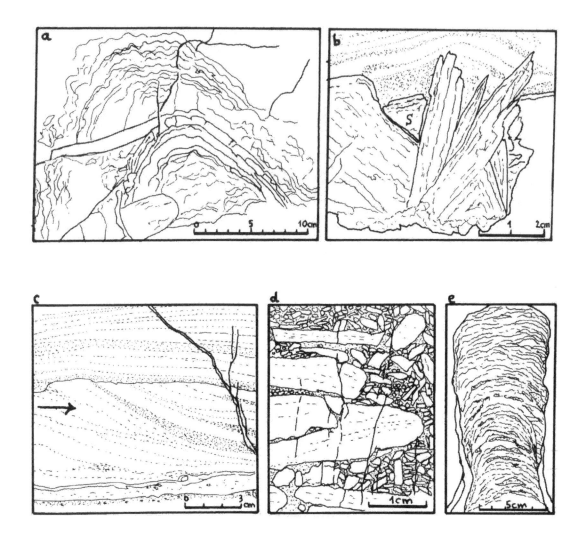

*Fig. 9.3*: a)–d): Rock strata from North Pole in NW Australia. The oldest terrestrial sedimentary rocks, these are embedded in vulcanites with a cooling-down age of 3500 MY. a) Silicified, arcade-like curved formations, probably a cutting face through a former chalky stromatolite. b) Heavy spar crystals (barite), which possess the crystal form of gypsum; therefore gypsum was replaced by barite (pseudomorphosis). The vertex S had broken from the left and fallen down. The gypsum crystals were covered by sand. c) Sandstone with typical discordant bedding. The lower stratum was formed into a ripple by the wave motion of the water, later partly eroded, and finally covered by another sand filling. The rock is completely silicified. The arrow indicates the flow direction of the water. d) The likewise silicified former lime sludge represents former tidal mud, which hardened, broke and was subsequently redeposited after only a short transport (little curved). e) Section through a current [extant] stromatolite from a supersaturated lagoon in west Australia. Note the typical arcade-like vault, developed through the carbonate secretions of cyanobacteria colonies (comp. Fig. 9.3a). (redrawn after Groves, Dunlop, and Buick 1981)

of the ancient landscape, which consisted mostly of a shallow, warm sea, interrupted by small and larger volcanic islands.

Another convincing stratum is added to the hornstones—a sulfate layer of heavy spar crystals up to 20 cm high (Fig. 9.3b). Such morphogeneses are known in present-day desert areas, where lagoons have evaporated. In such areas, we find gypsum crystals with the characteristic swallowtail feature with a similar spatial coordination and size of sulfate deposits. The transformation of gypsum to barite without changes in shape is known from geologically younger rocks. This is a process comparable to the silicification of wood. As barite crystals are less weatherable than gypsum crystals, we are fortunate again in the situation of discovering a preserved ancient deposit of warm lagoons. If we pay attention to the vertexes of these former gypsum crystals, which are partially broken off and lie between the crystal pillars (Fig. 9.3b), we can further develop the picture of a warm, shallow lagoon: The lagoon must have been submerged from a neighboring open sea during rare storms. This is supported by the cyclic sequence of various evaporation deposits. In addition, the sequence of unrounded sands with clay and silt rocks, combined with the characteristic ripple phenomena of mudflats, points to the existence of tides, as they are known from the interrelation between earth and moon (Figs. 9.3c, 9.3d).

The search for direct fossils of early life within these old sedimentary rocks was not successful. [Translator's note: Several microfossils have been found; however, the evidence—matching size, shape and biogenic carbon—is still only circumstantial. (Schopf 2009; De Gregorio et al. 2009; Brasier et al. 2006)] The structure of the stromatolites offers indirect evidence of life processes.

Additional evidence for early life is found in the study of the formation of the old sedimentary rocks.

## The Formation of Stromatolites from the Activity of Cyanobacteria

Studying extant stromatolite formations can be described as follows. Cyanobacteria colonies live in a shallow, warm, hypersaline and oxygen-deprived water body. Through photosynthesis they build all life substances from water and carbon dioxide utilizing the energy of light. Oxygen is released into the water as a byproduct. If oxygen were to stay in the water over an extended time, it would be poisonous for the cyanobacteria and destroy them. As the environment contains carbonate ions and dissolved calcium, lime is precipitating at the lower side of the bacteria colony. It is this lime that forms the stromatolites. The oxygen itself is not included into this calcification, but it obviously helps the chemical process of lime secretion to break through. However, oxygen bonds, for example ,to dissolved metal salts of a lower oxidation state.

In old Precambrian marine deposits, stromatolite forms are often found together with banded ironstones. This is a cm-interstratification of reddish siliceous rocks (lydites) with layers of hematite and bluish-gray hematite, a compound between oxygen and divalent iron to the insoluble trivalent iron oxide ($Fe_2O_3$). Slightly less frequent and deeper are layers of uranite, the compound of uranium and oxygen to form uranium oxide ($UO_2$). In other places the stromatolite layers are in interstratifications with sulfates, in compounds of calcium, sulfur and oxygen. Oxygen released as a waste product of photosynthesis obviously caused the formation of carbonates in sea water and was vonded as metal oxide or sulfate.

What do we discover here? The multitude of chemically developed sedimentary rocks could be formed only in the presence of oxygen. And according to extensive research and considerations, oxygen could develop only through life processes. And this happened at the time of the first rock formations on earth!

Another argument for the presence of early life in these oldest sedimentary rocks is the discovery of organic carbon, which is distinguished by its exceptional ratio between the carbon isotopes $^{13}$C and $^{12}$C. Such carbon deposits are the remains of biomass which was deposited after the death of organisms. This carbon is found either in form of the mineral *carbonate-hydroxylapatite*, which underwent little to medium metamorphosis (Hayes 1996), or in little globules of a coal-like substance of 0.01 mm diameter. Sometimes several of these globules are structured into a string or filament or in four clumped into tetrads in the sediment. (Groves, Dunlop, and Buick 1981)

## A Methodological Interjection

The appreciation of these rocks, e.g., in northwest Australia, is based on the principle of actualism (compare to Reif 2000; Gould 1965, 1987; Baker 1998; Anderson 2007; Dullo 1999), although they constitute the oldest sedimentary rocks found so far. The analysis of the microstructure of these rocks, combined with silicification and replacement by barite, is comparable to extant rock formation and sedimentation. A rejection of this actualism cannot be justified by these observational results (refer to Fig. 9.3). The statement (refer to Bosse 1995, 1999) that the first step would have been a hornlike compaction and crystallization that could not have happened before the Cenozoic, approximately from the Tertiary on, is in strong disagreement with the facts of barite crystallization in these sediments. The same laws of nature as today's must be expected for the chemical and physical processes of the earth's surface in earlier times. Otherwise the obvious comparability of the recent and old rocks would be highly unlikely incidents. However, what changed dramatically over time is the quality of *intensity of life processes* and, connected to this, serious transformations of the conditions for life. This shall be further investigated in the following.

## The Importance of Life Processes in the Transformation of Earth

If we look at the consequences that followed these early life processes, the story gets exciting and relevant at the same time. If we calculate the amount of oxygen needed for the rock formations (sulfates, ores, carbonates), the result is about 25 times more than today's free oxygen in the atmosphere and the oceans. This oxygen production by photosynthesis in combination with the depositing of organic and inorganic carbon led to a drastic reduction of the carbon dioxide content of the atmosphere, although carbon dioxide was continuously added into the atmosphere through the activity of volcanoes. With the reduction of this greenhouse gas, the global surface temperature decreased and the atmospheric surface pressure dropped to approximately today's values (chapter 8). What does this mean? By means of their excretions did early life itself create the conditions suitable for later, more complex and higher evolved forms of life? If we include the fact that during the geological history of earth, the sun's irradiation increased by a third and new greenhouse gases were constantly released into the atmosphere by volcanic processes, the enormous importance of these life processes needs to be admired even more. Without them, the earth would have an atmosphere increasingly unfavorable to life, comparable to Venus, where we find surface temperatures of approx. 500°C.

## Archaea (formerly known as Archae- or Archeobacteria)

After discussing cyanobacteria, which already mastered the biologically "modern" principle of photosynthesis despite their simple organization without nuclei (prokaryotes), we need to ask whether there were any simpler organisms before them. Today we know of simple bacterioides that survive without photosynthesis. They operate on so-called chemosynthesis. Ammonia bacteria that are, for example, used in sewage treatment plants combine nitrogen and hydrogen sulfite to ammonia utilizing sunlight. This releases elemental sulfur. Methane bacteria combine carbon dioxide and hydrogen sulfite to methane. They release sulfur dioxide and sulfur.

During the early times of the earth, conditions were probably excellent for such bacteria. Nitrogen, hydrogen sulfite and carbon dioxide were sufficiently present in water and air. In addition, gaseous oxygen was missing, which would have been a life-threatening poison for these bacteria. There is no convincing evidence for the presence of these bacteria. There are no specific deposits associated with them, either then or now. The only indications are frequent pyrite deposits in the oldest non-volcanic rocks, which suggest the release of elemental sulfur.

Dealing with sulfur compounds seems to have played a much more important role in earlier times. This is indicated through studies of today's archaea. In hot, oxygen-vacant volcanic water bodies, bacteria exist which reduce sulfates released by volcanic activity,

release hydrogen sulfite and oxygen, and gain energy for their life by doing this. Purple sulfur bacteria on the other side, combine hydrogen sulfite and carbon dioxide utilizing sunlight and turn them into life-building glucose. As products which fall out of the life process, sulfur and water are produced.

The pyrite content of very old rocks (older than 2000 MY), which settled down in water is striking. It is possible to assume that the released sulfur combined with dissolved iron in seawater, formed pyrite, and, due to the absence of oxygen, did not disintegrate.

These indications only make it clear that photosynthesizing cyanobacteria were surely not the first organisms on earth. This means that *there was life before the emergence of the first rocks on earth*. We have come one important step closer to the question of primacy of life or dead rock.

## The First Turnaround: Free Oxygen in the Atmosphere

If we stay with the sedimentary rocks that formed through the course of geological time, we find a striking change starting at about 2000 MYA. There are no more marine iron deposits (banded iron formations, also known as banded ironstone formations or BIFs). Instead we find the first continental deposits in a red color. This means that dissolved [divalent] iron from crystalline rocks was already on the mainland and combined with oxygen to form red sediment. These can be sandstones, conglomerates or brecciae. From this process, we can surmise that there was no iron left in the oceans to allow the formation of banded iron formations. The turnaround at 2000 MYA is therefore the beginning of an oxygen oversaturation of the world oceans from the seafloor to the surface, as well as the beginning of oxygen being released into the atmosphere. This does not mean that the atmosphere held on to this oxygen; instead it was immediately reabsorbed to form continental trivalent iron and uranium ores (pitchblende, uraninite). There are also no more marine uranite formations after this time.

Another first from this period is the appearance of deposits from arctic climates. Moraines and scratch marks on rounded rock humps document the first glaciations on both poles of that time. In addition, *Warfen deposits* appear. These are very fine, rhythmically layered clay deposits that record the changes between summer and winter in arctic climates. (Stanley 1994, 266) For the first time, global temperature must have reached values that were in some places comparable to today's climates. (Canfield 2005)

### Steps in the Development of Life

Let us summarize: The first documents of life processes were found in the oldest sedimentary rocks around 4000 MYA. Single-celled organisms, preserved as microfossils were found in sediments 3500 MY of age. In slightly younger rocks, small globules were lined up into strings; these are called pre-algae but they are still considered to be single-celled organisms (protists). In the Gunflint banded iron formation, dated to be 2000 MY old, with the help of electron microscopes, a well-preserved series of fossil single-celled organisms was found, which resemble iron bacteria, cyanobacteria and simple algae. All of them are unicellular organisms without nuclei. Documentation of the appearance of *eukaryotes*, unicellular organisms with nuclei, is found in 1500 MY old sediments. (Vidal 1984; Bengtson et al. 2009; Rasmussen et al. 2008) A striking difference is a several times increased size and a more differentiated inner structure, which is recognizable in a

few extremely well preserved microfossils. The conclusion that these were probably eukaryotes suggests itself immediately, as the oxygen needed by eukaryotes was already present in small amounts in the environment of these water organisms. Current research (Martin 2005; Martin and Müller 1998) establishes the hypothesis that first eukaryotes originated from a symbiosis between archaea and methane bacteria. According to this hypothesis, eukaryotes are not a further development of "modern" photosynthesizing cyanobacteria, but a brilliant symbiosis of two very different and very primeval groups of bacteria.

The next major, and well-documented, step is the appearance of real multicellular organisms. The spectrum of forms and shapes of these small multicellular organisms (size range in millimeters) within limestones of approx. 900 MY of age is still very simple. Common forms are tetrad four-cell states, spheres and elongated forms with a thickening on one side. (Vidal 1984)

If we take into account the long time of 1400 MY, during which the atmosphere had a certain content of oxygen, it is astonishing to see how little organisms differentiated. (Butterfield 2007) This was also a period of rest in geological aspects—in terms of abnormal orogenesis, volcanism and sedimentation. However, it is reasonable to assume steady plate tectonic processes, including the enlargement or reduction of oceans and the relative movements of continents. The times around 1600 MY, 1100 MY, and 700 MY, for example, were characterized by a strong alternation in phenomena from compression to dilatation, the so-called stage of Pangaea, in which continents unified to form one large megacontinent (Rodinia) surrounded by a megaocean.

## The Ediacara Event

In the youngest rocks of the Precambrian in various locations of the earth, the first complex, large multicellular organisms appear. This group of fossils was named Ediacara fossils after the first location where they were found, Ediacara in Australia. The most interesting fact about these strictly rhythmically and geometrically structured organisms up to 10 cm in size is their lack of hard structures, causing a high uncertainty in interpretation as to whether they are animals or plants. Within some of these soft-tissued organisms, geochemical traces of chlorophyll were found; other structures clearly resemble animals. Researchers assume an intensive symbiosis of more animal-like organisms with photosynthesizing plant-likes. Fossils were found in fine clay shale, together with warm water sulfates and carbonates. This indicates that the habitat of these organisms was in shallow waters within the equatorial climatic band. If we plot the locations of discovered fossils (Australia, China, Norway, Canada) into a paleocontinental world map showing the world 650 to 700 MYA, we find this assumption confirmed (Fig. 9.4).

The richness of forms (Fig. 9.5) of typical Ediacara fossils (Narbonne 2005) animates students to speculate in many different ways. We cannot help thinking that this group of organisms does not yet allow for a distinction between the realm of plants and the realm of animals of modern geological times. We discover again that universal forms first develop and then differentiate in a second step. (Knoll, Walter et al. 2006)

## The Appearance of the Realm of Animals

The turnaround from the Precambrian to the Paleozoic is most of all characterized by the landing of a multitude of animal fossils. Although the most recent research results argue no distinct separation between the extinction of the Ediacara organisms and the appearance of real animals (Gaidos et al. 2007; Jensen, Gehling, and Droser 1998; Shen, Zhang, and Hoffman 2008; Budd 2008), it is still striking how, compared to the previous hundreds of millions of years, all of a sudden a huge variety of animal blueprints appear, animals that we still find today. (Gould 1989; Levinton 2008; Levinton, Dubb, and Wray 2004; Levinton 1992; McMenamin 2000, 1987)

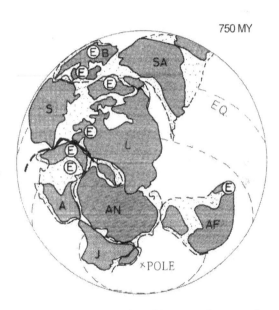

750 MY

*Fig. 9.4:* Paleocontinental world map at the time of the late Proterozoic megacontinent at the transition to the Vendian (750 MYA). The continents Australia (A), Africa (AF) and South America (SA) are shown with their flooded parts (dotted). Flooded regions of the continents Baltica (= Europe, B), Siberia (S), Laurentia (= North America, L), India (J) and Antarctica (AN) are not shown. The bold-dashed line shows the mid-ocean ridge which separated the group Australia-Antarctica-India from Laurentia during the Vendian and, as a consequence of this, pushed these continents and Africa towards South America (and closed off the old Atlantic Ocean). The locations of Ediacara fossils (E) fall all within the tropic and subtropic bands of that time and at the margins of shallowly-flooded continental areas. (according to Dalziel 1995, modified and supplemented)

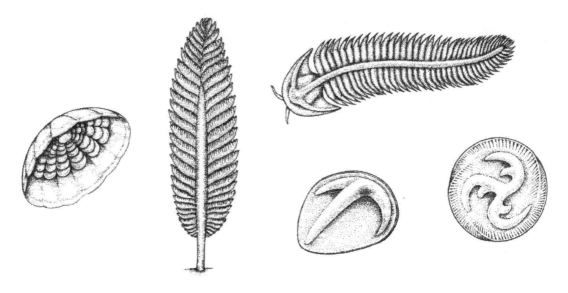

*Fig. 9.5:* Typical diversity of shapes within Ediacara fossils, found in south Australia in the Adelaide Basin. The gesture of shaping [Gestalt] shows relatedness to plants, open to their environment, as well as to animals, which form an inner space. Characteristic are the geometrical symmetry and the jointed, rhythmical structure, respectively. (from Cloud 1983)

Within a short interval of time of just a few million years, 19 animal phyla developed, from which 6 are already extinct (see Figs. 9.6 and 9.7). The sensational diversity in shape of the fossilized animals in the *Burgess Shales* in the middle Cambrian of western Canada discloses most of all creative forms of segmented animals. Similar to the discoveries of the Ediacara organisms, a couple of fossils which were attached to the muddy soil show a plant-like structure. The reconstruction of the habitats for that time shows once again a shallow, warm and oxygen-rich water body close to the coast (see Fig. 9.8).

It is astonishing that an accompanying flora in the modern sense is still missing. The Burgess animals fed on algae or lived in symbiosis with photosynthesizing bacteria. There were, however, also the first predatory animals with armored exoskeletons and gripping devices. As an exceptional case, please note Fig. 9.7k: This is the oldest specimen of a chordate announcing the emergence of a spinal column! If one searches for an understanding of this biologically-rich development, it pays off to investigate the conditions of life during that time. (Marshall 2006) When the last large Precambrian glaciation—accompanied by the extinction of the Ediacara—was over, the continents were loosely distributed around the Equator (Fig. 9.9). Now we can return to the 10th grade earth science main lesson block and out discussion of the ocean currents. At the diminishing poles, surface water sinks into the depth and is driven towards the Equator. Slowly deflected by the mid-ocean ridge, the nutrient-rich deep sea water rises at a multitude of coasts around the Equator and where an appropriate influx of sediments from the mountains is added, ideal conditions are created for the development of aquatic life. It is said that, if we were to try to come up with ideal conditions for the development of the Cambrian life, we would arrange oceans and continents just the way as they are shown

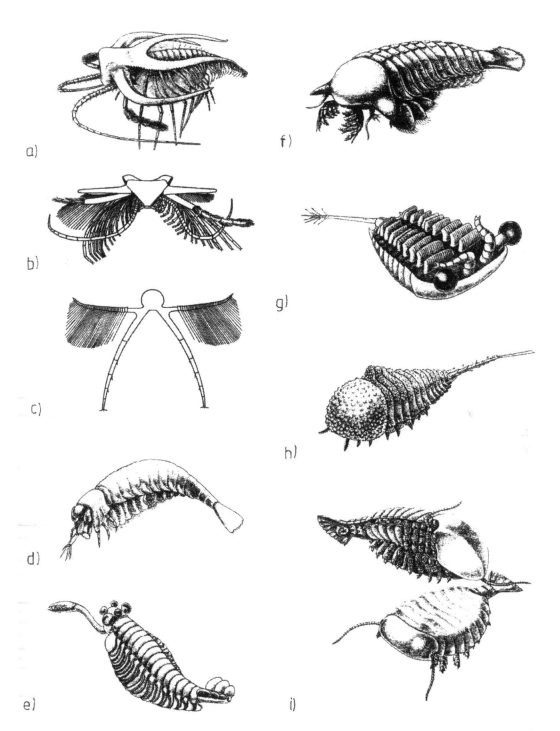

*Fig. 9.6:* Typical animal shapes of the Cambrian Burgess formation in Canada. These reconstructions provide a picture of the highly differentiated grouping of these inhabitants of shallow waters. a)–i) are extraordinary arthropods: a) side view of *Marella*, b) frontal view of *Marella*, c) symmetric design of *Marella*, on top the two gill branches, below the two leg branches, d) the arthropod *Yohoia*, e) *Opabinia* with five eyes and a long proboscis, f) *Nectocaris*: In the front rather an arthropod, in the back rather a chordate, g) *Sarotrocerus*, swimming on the back, with two ball-shaped eyes, h) *Habelia*, a blind arthropod, i) two views of *Sidneyia*, with two small eyes, antennae and a multitude of limbs. (from Gould 1989)

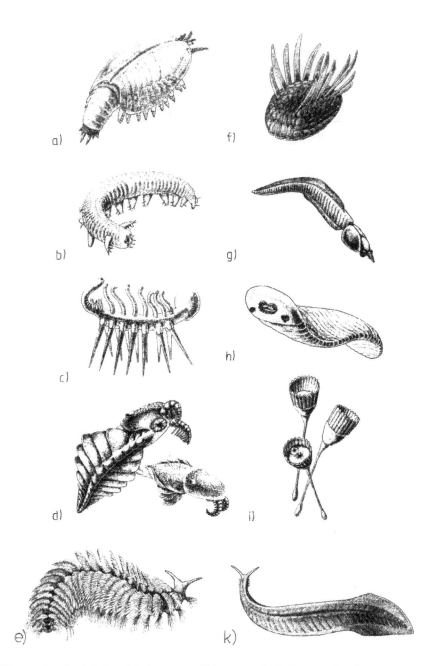

*Fig. 9.7:* New exotic animal phyla and the basic animal blueprints of today's species found only in Cambrian strata. a) *Canadaspis,* a real crustacean with the blueprint of a lobster; b) *Aysheaia,* a mixture between an arthropod and an annelid, still around today in the very similar velvet worms (*Onychophora*); c) *Hallucigenia,* maybe an extinct phylum, the small animal stood on seven pair of stilts; d) two different *Anomalocaris,* presumably good swimmers and, compared to the other animals, of outstandingly large body structure, another extinct animal phylum; e) *Canadia,* a precursor of the polychaeta, which are today highly differentiated; f) *Wiwaxia* crawling on the bottom of the sea, its chewing apparatus similar to mollusks, otherwise more similar to arthropods or annelids, according to recent research most probably a polychaete; g) *Nectocaris* shows similarity to chordate in its hind end but the head is more arthropod-like; h) the jelly-like swimmer *Odontogriphus* possesses two round sense organs on its underside and a toothed mouth, another extinct phylum; i) the stemmed *Dinomischus* was sessile and resembles current bryozoans but shows significant differences justifying another extinct animal phylum; k) *Pikaia* is the oldest chordate ever found and as such the basic form of all vertebrates. (from Gould 1989)

*Fig. 9.8:* Reconstruction of the environment of the Burgess fauna, consisting of coastal sludge layers in a water depth of approximately 500 m. From Fig. 9.6 and 9.7 there are: *Marella* (15), *Yohoia* (11), *Opabinia* (8), *Canadaspis* (12), *Aysheaia* (5), *Anomalocaris* (24), *Wiwaxia* (23), *Dinomischus* (9) and *Pikaia* (upper right, small), sessile sponges (25, 22), sea lilies (*Crinoids,* 21) and corals. (from Gould 1989, image width approx. 60 cm)

in Figure 9.9. The climate was favorable and the salinity of the oceans was in balance. The only situations hostile to life were on land and in the air. Due to the ozone layer not yet established in the stratosphere, life was possible only in water, where it was protected from UV radiation.

### The Differentiation of Vertebrates

After all invertebrate animal phyla, including the first representatives of ancient vertebrates, spread out globally within the shallow seas, the following time of the Paleozoic and Mesozoic were characterized by the development of forms from tetrapods to mammals. A reference very suitable for teaching in grade 12 is *Paleontology as Anthropology* (Suchantke 1967). A good presentation of the sequence of life forms

and the corresponding environment can also be found in the book *Earth System History* (Stanley 2008). I will highlight only two aspects which may be of significant interest to the students.

### The Puzzle over the Coelacanth

The first fossil documents of the formation of four bony limbs were preserved inside of Devonian slate (approx. 410 MYA). Coelacanths and lungfishes (*Dipnoi*) are considered to be the relatives to the ancestors of amphibians and, as such, all higher vertebrates. Although they are quite lumpy in form, the bones of the new limbs of the coelacanths show a clear pattern of 1-2-3-4-5 when counted from the inside out, which corresponds to the pattern in amphibians, reptiles and mammals.

This animal group was long believed to be extinct, and it was unclear what sense these new bony structures served. According to the neo-Darwinian way of thinking, it was believed the coelacanths lived in mudflats which are formed by tidal movements. Fishes could miss the outgoing tide and die in the mud. It is believed that an accidental mutation causing this new ossification provided a survival advantage for these fishes. The four strong lower fins would have provided them with the ability to crawl back into the water and thus allow them to forage in the nutrient-rich mudflats longer than other competing species.

After finding and filming still-extant specimens of coelacanths off the steep coast of Madagascar in the 1960s, a completely different way of living was discovered. (Thomson 1991) These fish, up to 1 m long, are nocturnal predators at the steep slopes in 100 to 300 m water depth. When the animals were provoked by electromagnetic waves, the males moved their fins in such a way that the presence of the described bony structure made sense: They moved all four fins and rotated in place around their longitudinal axis, while keeping their heads always perpendicular to the rocks, regardless of the orientation of the rocky surface in space. The researchers had to realize that the limb bones that were so important in evolution offered the opportunity for a very unusual courtship dance in these rare fishes! This example shows clearly how wishful thinking and real nature observation can be far apart when wishful thinking wants to prove a perhaps unrealistic theory. More information about these amazing "living fossils" can be found in: Heemstra et al. 2006; Fricke 2001; Fricke et al. 2000; Fricke and Hissmann 2000; Fricke et al. 1987.

## Reptiles of the Paleozoic

Land and even the air is vitalized over the following course of the development from amphibians to reptiles. The landmasses become the stage of vertebrate life for the first time. If we visualize all the habitats that were colonized by Permian reptiles, it is fair to say that every spot of landscape became inhabitable for animals. Even the extreme climatic zones of deserts and polar regions were colonized. Even the airspace was conquered by pterosaurians before real birds appeared. Even the oceans were not neglected: The strongest predators of the oceans were reptiles and not fishes. It is therefore possible to state that the evolution of animals came to a preliminary end in the Permian. Emancipation progressed to the point that reptiles found a niche everywhere on earth.

## Unfolding and Diversification of Plants

Surprisingly, the systematic diversification in plants occurred much later than in animals, and therefore it is important that the class teacher in middle school does not raise the imagination that plants evolved before animals. Unfortunately, this misrepresentation happens quite often in geology blocks in grade 6 when the Paleozoic is described as the world of plants and the Mesozoic as the world of animals.

The world of plants diversified tremendously with the development of the first land [terrestrial] plants. Structures have been found in the strata of the Ordovician (approx. 450 MYA) that resemble the spores of today's seedless plants. Unambiguous land plants were found in the Silurian. Plant chaff was found in coastal swallow water deposits in the river deltas, which allowed for the reconstruction of vascular plants that showed a bushy form developed from a dichotomous branching of the shoot. The principle of uplift/erection from solid ground into the airspace was achieved.

Massive diversification of land plants did not happen before the Devonian, about 400 MYA. These were fern- and horsetail-like plants, as well as Lycopodiales showing real root organs. Their habitat was mainly in the riparian zone and in swamps. In contrast to their relatives today, these plants reached an enormous height, which even increased in the following Carboniferous. The special feature during the Devonian was therefore a tremendous vitality, a rampantly growing, relentless proliferation in biomass.

If we plot the position of oceans and continents during the lower Devonian (Fig. 9.10), we realize immediately a significant difference in the Southern polar area when compared to Figure 9.9. Instead of an ocean that would have allowed the down-welling movement of surface water, we now have the super-continent Gondwana. The continents of the Northern hemisphere were distributed in such a way that the still existing down-welling currents in the Northern polar region were only surfacing at the western coast of Laurentia as they were partially limited through the distribution of the mid-ocean ridges. The conditions for life at the transition zone between deep-sea and shelf sea became less favorable compared to the conditions in the Cambrian. Marine extinctions at 510 MYA (end of the Cambrian) and 440 MYA (end of the Ordovician) were caused by these changes and massive glaciations during the same time. Coastal lowlands were now predominantly in subtropical and temperate climate zones—ideal conditions for the colonization of land by plants and animals. A slightly higher sea level also contributed to an increase in area of swamps in the transition zone between land and sea. Compared to the late Cambrian glaciations, the global climate

was warmer, which was caused not least by the intensive activities of marine animals. The paleocontinental position during the Devonian is a second example students can experience of how the varying face of the earth makes sense: It provides supporting conditions for new steps in the development of life.

## About the Formation of Coal during the Carboniferous

Plant vitality reached a geological climax during the middle and upper Carboniferous. Coal fields are the consequence of intensive life and die-off processes. When realizing that several hundred meters of plant chaff are needed to produce 1 m of coal seam, we understand this immense plant vitality that was always close to death. When consulting Figure 9.11, we can repeat the conditions that provided these plants with such a vitality.

Because of the upwelling of nutrient-rich fresh deep-sea water, life conditions were still good at the west coast of Laurentia and northern South America. The salinity of the oceans became one-sided for the first time due to an almost completely missing global east-west current of the water. In the Northern hemisphere, the conditions for life on land were very good. Many coastal areas were shallow but received nutrient-rich rubble from the newly developing high mountains of the Urals, Altais and Appalachians. The climate was relatively humid for sure in the subtropical regions but maybe also in the temperate zones up to higher latitudes, due to the constant alternation between a narrow ocean and a not too large land mass.

Another crucial aspect for the formation of coal deposits were cyclic transgressions and regressions caused by minor fluctuations of the sea level. This was most probably caused by the cyclic advances and regressions of large inland glaciers on the southern Gondwana continent in the Southern hemisphere. This cycle was most probably also driven by the cosmic rhythms of the motion interrelations between earth and sun as documented for the latest, most current ice age. When discussing this topic with the class, we can successfully go back to the astronomy block of grade 11. The question of why the global climate got colder during a time of plant blossoming should be easily understandable. Excessive deposition of carbon in the earth's interior allowed the atmospheric carbon dioxide content to decrease drastically (comp. Fig. 8.2 in chapter 8).

## The Global Crisis at the End of the Paleozoic

A further study of the paleocontinental world maps (Figs. 9.12 and 9.14) shall help us to understand the conditions of that time. The position of the unified Pangaea continent made the exchange of ocean water impossible, since an east-west circulation could not happen. The land climate was very extreme due to the large connected landmasses causing hot summers and cold winters. Mighty glaciers developed near the poles. With the exception of the North Pole region, steep coasts dominated the land-water interface. It is questionable if the Tethys Sea was that open towards the old Pacific Ocean,

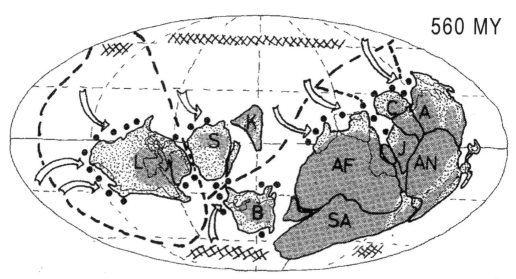

560 MY

*Fig. 9.9*: Paleocontinental world map at the transition between the Vendian and Cambrian (560 MYA). The Northern and Southern polar seas possibly had large downwelling sites of surface water. As a consequence, strong upwelling sites would have occurred along many coasts in the tropical and subtropical climate zones. (according to Bambach, Scotese, and Ziegler 1980; Stanley 2008, supplemented)

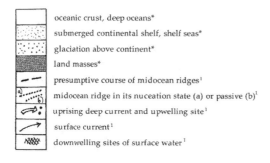

oceanic crust, deep oceans*

submerged continental shelf, shelf seas*

glaciation above continent*

land masses*

presumptive course of midocean ridges[1]

midocean ridge in its nuceation state (a) or passive (b)[1]

uprising deep current and upwelling site[1]

surface current[1]

downwelling sites of surface water[1]

LEGEND Figs. 9.9–9.17: The information about ocean currents includes uncertainties and is meant as general indications. Continental shelf: L = Laurentia (North America, Greenland), B = Baltica (Europe), S = Siberia (Northeast Asia, Siberia), K = Kazakhstan, C = China, A = Australia and New Guinea, J = India, AN = Antarctica, AF = Africa and Madagascar, SA = South America, SA+AF+AN+J+A+C = Gondwana continent, Gondwanaland. (according to Bambach, Scotese, and Ziegler 1980; Howarth 1981; Owen 1983; Stanley 2008)

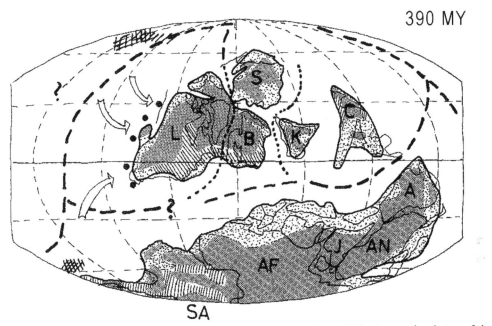

*Fig. 9.10:* Paleocontinental world map of the upper lower Devonian (390 MYA). Due to the closing of the old Atlantic Ocean, Laurentia and Baltica united. Gondwanaland migrated to the South Pole, resulting in a subtropical east-west sea in the Southern hemisphere. Down- and upwelling movements were reduced to the old Pacific Ocean. Shallow-water fauna provinces in Laurentia and South America are diagonally shaded. (according to Stanley 2008, supplemented)

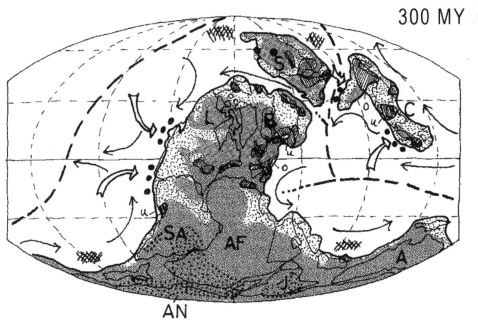

*Fig. 9.11:* Paleocontinental world map of the upper Carboniferous (300 MYA). The large Gondwana continent disables an east-west circulation of ocean water besides a small gap between Baltica and Siberia, shortly before the formation of the Ural mountain range. Today's coal deposits are shown as diagonally shaded areas in the lower Carboniferous (U = Mississippian) and upper Carboniferous (O = Pennsylvanian). The climate was very different in both hemispheres: In the south there was a far north-reaching ice age; the Northern hemisphere was humid and cool. (according to Stanley 2008, supplemented)

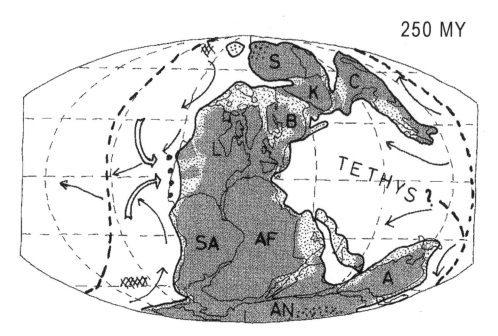

*Fig. 9.12:* Paleocontinental world map of the upper Permian (250 MYA). The continent Pangaea made a continuous east-west circulation of ocean water impossible. As the water stagnated within the very large and deep old Pacific Ocean, only a small part of the continental blocks was submerged. It is questionable whether the Tethys was that large (comp. Fig. 9.13). (according to Stanley 2008, supplemented)

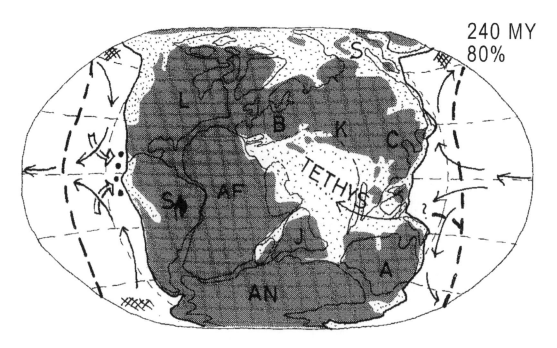

*Fig. 9.13:* Paleocontinental world map at the transition between Permian-Triassic (240 MYA) on a globe that is 20% smaller than today's globe. Size and position of the Tethys and the Northern polar seas match much better to the geological findings. Compared to Fig. 9.12, the continents form an even more compact landmass, which caused more severe climatic extremes. (according to Howarth 1981; Owen 1983, supplemented)

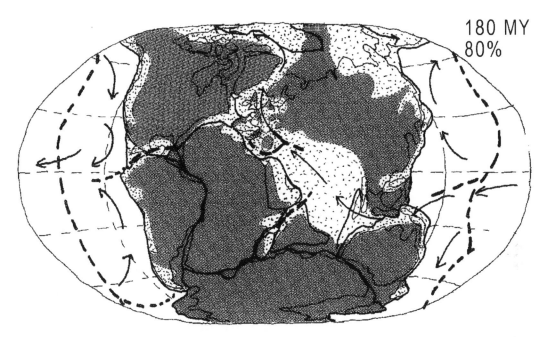

*Fig. 9.14:* Paleocontinental world map during the lower Jurassic (180 MYA) on a globe that is 20% smaller than today's globe. A comparison to Fig. 9.16 shows the absence of a South polar ocean, only a small opening of the Tethys at its eastern margin, and the lack of a North polar sea. (according to Owen 1983, supplemented)

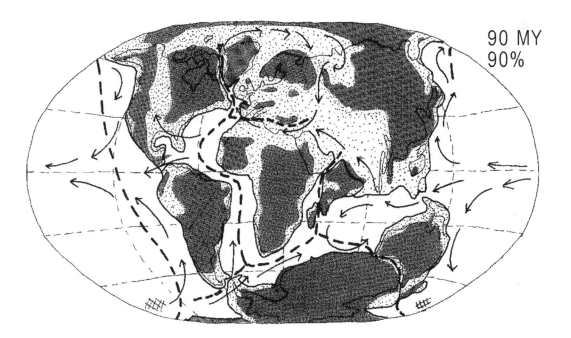

*Fig. 9.15*: Paleocontinental world map during the Upper Cretaceous (90 MYA) on a globe that is 10% smaller than today's globe. In comparison with Fig. 9.17, the oceanic east-west circulation is intensified within the Tethys of the Northern hemisphere and in the Indian Ocean in the Southern hemisphere, due mainly to the fact that an Antarctic southern current was not possible. (according to Owen 1983, supplemented)

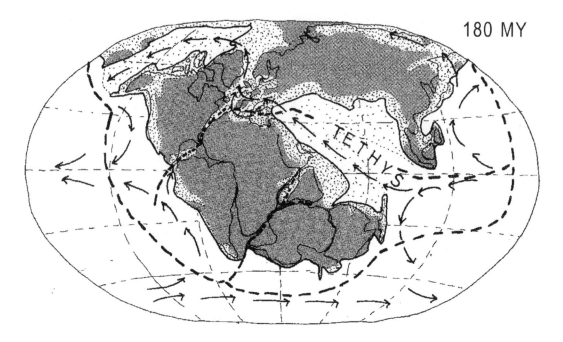

*Fig. 9.16:* Paleocontinental world map of the Lower Jurassic (180 MYA). Pangaea broke apart at the Tethys sea, and the continental shelves were more and more submerged, especially in the Tethys area. It is questionable whether such a large South Pole ocean was present (comp. Fig. 9.14). (according to Howarth 1981, supplemented)

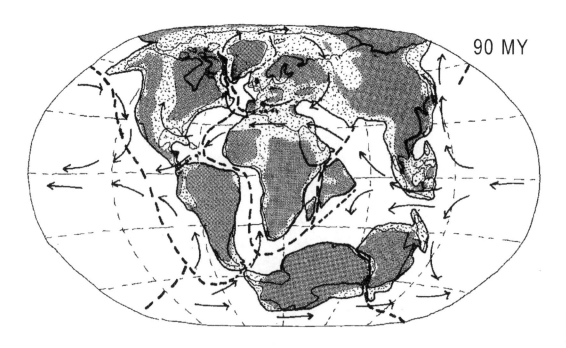

*Fig. 9.17:* Paleocontinental world map of the Lower Cretaceous (90 MYA). (according to Owen 1983, supplemented)

as the paleocontinental world map in Figure 9.15 shows a much smaller Tethys (about Fig. 9.15, refer also to the following excursus). The entire configuration is characterized by an extreme *polarization*. In addition, this situation continued for more than 50 MY, which added *solidification* to the polarity.

Estimates assume that up to 90% of all animal and plant species went extinct within a timeframe of 40 MY. Especially hurt were the Actinopterygii and Sarcopterygii within the fishes, Mesosauria and Pelycosauria within the reptiles, Cordaitales and Glossopteridales within plants. When searching for reasons for this and earlier Paleozoic extinctions, the cooling of the world oceans seems to play a major role. (Stanley 1985; Stanley 1987; Stanley and Campbell 1981; Stanley and Powell 2003; Stanley and Yang 1994) With the exception of the crisis at the end of the Ordovician, the extinction events of 500 MYA, 440 MYA, 400 MYA, 340 MYA, 280 MYA and 230 MYA have not only been associated with a cooling of the world oceans but also to a drawdown of the sea level, which meant a massive shrinking of shelf seas. The synergy of the position and form of the world's oceans and the global climate constituted periodically deteriorating conditions for aquatic and terrestrial life. There is a striking periodicity to these extinction events of approximately 60 MY, which continues into the Mesozoic (refer to Fig. 8.2 in chapter 8).

## Excursus: Constant-sized Earth or Expansion Since the Mesozoic?

The research of Owen (1981, 1983) shall be discussed as an example of questioning long-held assumptions in the field of Geology. When reconstructing the sizes of oceans and the positions of continents from present time backwards until the Permian, a much larger ocean emerged within the area of the Tethys and the inlet between North America and Siberia than what has been backed up by geological evidence. Owen made these apparent "holes" disappear, by assuming that the earth's radius was 20% smaller at the time of the transition between the Paleozoic to the Mesozoic than it is today, and that it has enlarged continuously since then. Owen's calculations about the ocean and continental positions are shown in Figures 9.13, 9.14 and 9.15. In comparison to Figure 9.12, Figure 9.13 shows an even more extreme polarization of climatic conditions. In Figure 9.14, Antarctica was where it should be according to paleoclimatic evidence. In this context, Figure 9.16 is problematic. There is not much difference between Figures 9.15 and 9.17. In Owen's reconstruction, the east-west circulations north and south of Africa are stronger, and also the North hemispheric circulation around Greenland and the Baltics can be expected to have been stronger.

Which inner reasoning does Owen present to support the hypothesis of an expansion of earth? He proposes constitutional changes in the earth's outer nucleus in correspondence with a shifting of forces in the solar system. This would also explain why the moon without a corresponding nucleus obviously did not expand.

I want to throw in just a spotlight from the presentation of Rudolf Steiner's about the development of the earth as it was summarized in the teacher's circular from October 1993. (Verhulst 1993) According to Verhulst's interpretation of Steiner's descriptions, the material earth should show an expansion, and there should be something like a gap or pocket in the area of the North Atlantic and Indian Oceans due to the delayed compaction of "Lemuria" and "Atlantis." Are Owen's reconstructions the natural scientist's counterpiece to the statement of the spiritual scientist?

I consider this parallelism to be problematic. On the contrary, the theory of a periodic expansion and compression of the earth's radius in the context of material changes in the earth's nucleus and mantle seems to be more plausible, as such pulsating phenomena of constriction and expansion occur in many areas. It also seems plausible that a maximum in contraction coincided with the appearance of a united world continent.

## The First Flowering Plants
## and the Propagation of Mammals

After the breakup of Pangaea, life conditions in water and on land changed such that most animal phyla experienced a renaissance, especially ammonites, bivalves and reptiles. In the kingdom of plants, this new impulse occurred during the Lower Cretaceous (130 MYA). In the area which is today's Indonesia, the first angiosperm flowering plants diversified as magnolia-like plants. Soon followed their triumphant advance along the coasts of the Tethys. Simultaneously with the appearance of colored flowers, the first mammals developed. After the extinction of dinosaurs during the Cretaceous crisis, mammals developed through all habitats on earth, just as reptiles had done at the end of the Paleozoic. It is interesting that the sophisticated opening of modern flowering plants towards the cosmos and the autonomy of animals [Rosslenbroich 2006, 2009—tr.] in terms of their heat balance and embryonic environment occurred at the same time.

The paleontological world maps in Figures 9.16 and 9.17 and the maps by Owen (Figs. 9.14 and 9.15), respectively, illustrate how the tearing apart of Pangaea

restored the oceanic circulations and put an end to extreme values in salinity. In the area of the Tethys and southern Indian Ocean, a climate similar to today's Atlanto-European climate developed. It was characterized by a rhythmical alternation between warmth and cold, and periods of fair weather and precipitation. It is certain that the global temperature was much higher than today's (comp. Fig. 8.2 in chapter 8) and without glaciations. The cooling that started in the Middle Cretaceous, and was accompanied by a pronounced decline of the atmospheric carbon dioxide content, was caused by the blooming of plants on all continents, as there were practically no landmasses within the deserts zones (comp. Figs. 9.15 and 9.17, respectively).

## Indications for Dating
## the Age of Atlantis

Steiner characterized the downfall events of the Lemurian Age as a catastrophe of fire in contrast to the collapse of Atlantis, which was characterized by flooding. (Steiner 1910a, chapter "Man and the Evolution of the World" *[Die Weltentwicklung und der Mensch]* (Steiner 1910a) The middle period of the Lemurian catastrophe was accompanied by the separation events between earth and moon. Earlier in this chapter, I wrote that the discharge of the moon can be dated to the earliest Proterozoic. I would suggest the following dating: The timeframe between the discharge of the moon and the beginning of the Atlantic era can be associated with the Proterozoic. Heat processes also dominated geological events. Just think about the hot early atmosphere and massive magmatic and volcanic processes that led to the formation of continents of sizes approximate to today's.

The Paleozoic, Mesozoic and Cenozoic can be described as a non-stop rising and falling of the water level, combined with massive biological death processes (comp. Fig. 8.2 in chapter 8). In the lecture cycle: *The Mission of Individual Folk-Souls in Connection with Germanic-Scandinavian Mythology*, Steiner describes the geographic position of the old Atlantis:

> Let us clairvoyantly observe the old continent of Atlantis, which must be sought where the Atlantic Ocean now lies, between Africa and Europe on the one side and America on the other. This continent was encircled by a sort of warm stream, a stream which clairvoyant consciousness reveals that, strange as it may sound, flowed upwards from the south, through Baffin Bay, towards the north of Greenland, encircling it and then, flowing over to the east, gradually cooled down; then, at a time before Siberia and Russia had risen to the surface, it flowed down near the Ural mountains, turned, touched the eastern Carpathians, flowed into the region occupied by the present Sahara, and finally streamed towards the Atlantic Ocean near the Bay of Biscay; so that it flowed in a perfectly unbroken stream. You will understand that only the remnants of this stream still remain. This is the Gulf Stream, which at that time encircled the Atlantean continent. (Steiner 1910b, 10th lecture on June 16, 1910, page 176f, 1929; authorized translation edited by H. Collison, Steiner 1989)

If we compare this citation with the paleocontinental reconstruction of the earth during the Upper Cretaceous (comp. Fig. 9.15), a surprising coincidence appears. The circumpolar current discussed by Steiner is actually reconstructible. The Atlantic living space at that time disintegrated into several smaller and larger islands. The middle of this group of islands is actually where today's North Atlantic Ocean is.

It can therefore be postulated that the Atlantic sinking events were already on their way during the Cretaceous and reached their conclusion at the end of the ice age in the fall of the Atlantic culture as described by Plato. Based on these considerations, the Atlantic Age covered the geological time between the beginnings of the Paleozoic until the end of the ice age. The crisis of solidification during the Permian would therefore not be attributed to the Lemurian catastrophe but would have the character of a memory of the Lemurian catastrophe. This hypothesis, however, is contradictory to Steiner's indications during the conferences. (Steiner 1924b, from September 25 and 26, 1919, translated as *Faculty Meetings with Rudolf Steiner,* Steiner 1998) A more detailed investigation is required to evaluate this temporal hypothesis. The process character portrayed by the spiritual scientist would have to be compared to the course of geological events.

## Anthropology

### *General Indications*

This is not the appripriate place to describe the paleontological sequence from the Cretaceous to the Pleistocene glaciations. Rather, in this class, we could discuss the diversification of mammals, the extinction of dinosaurs, the Cretaceous crisis, and the shaping of the modern fauna. Excellent literature is available (e.g., Stanley 2008). In further presentations we want to focus on paleoanthropology and highlight aspects

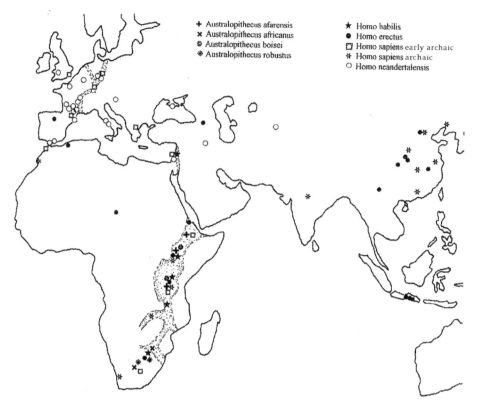

*Fig. 9.18:* The most important discovery sites of ancient humans, old humans, and pre-humans on the African-Eurasian landmass. (according to Johanson, Johansen, and Edgar 1994; and Bromage and Schrenk 1999)

that are important when discussing this topic in class. The following books provide a good overview of the history of discoveries and their scientific problems: *The Origin of Modern Humans* (Lewin 1993), *Origins Reconsidered: In Search of What Makes Us Human* (Leakey and Lewin 1992), *African Biogeography, Climate Change, & Human Evolution* (Bromage and Schrenk 1999), and *The Fossil Trail: How We Know What We Think We Know about Human Evolution* (Tattersall 2009). An excellent photographic documentation of the most important bone and skull finds is *Ancestors: In Search of Human Origins* (Johanson, Johansen, and Edgar 1994), with its more recent companion: *Lucy's Legacy: the Quest for Human Origins* [Johanson and Wong, 2009—tr.].

As species determination is very difficult in Hominids, various versions of Hominid systematic exist. The text presented in this curricula uses a widely accepted taxonomy of 1998.

### *The First Achievement: The Upright Stance*

When observing the skull forms of ancient humans, we can understand the scientists when they speak about a transition between apes and man (refer to Figs. 9.21 and 9.25). If we compare the entire skeletons, however, we find explicitly a clear difference between all apes and all Hominids. All Hominid skeletons show an upright stance, which once established never disappeared. It is a good idea to review the anthropology main lessons of grades 8 and 9, when the class acquired a detailed knowledge about the human skeleton. The principle of uplift and erection was followed by the elastic arch of the foot, the special construction of the knee, the form of the pelvis with its angular relationships between the spinal column and the thigh-bones, and the sweep of the spinal column up to the position of the skull in its unstable equilibrium above the body's center of gravity.

It is possible to divide the skeleton finds of the upright walking creatures into two groups (refer to tables A and B in the appendix of this chapter). Preglaciation finds (4 to 2 MYA) are restricted to the African continent, or more specific to the surroundings of the Central African rift valley system (Fig. 9.18). These Hominids are called *Australopithecines*. (This name should be changed in the near future as these creatures were definitely not "Southern Apes" but upright walking beings.) The skeleton finds of the ice age (2 to 0.1 MYA) are called *Homo* and are at first restricted to the "Old World," and occur from 0.4 MYA on also in the "New World."

In addition to these old African bone finds, corresponding footprints were found (Fig. 9.19). A good example are the 3.5 MY old footprints found near Laetoli inside the African rift valley in 1978. (Agnew and

Demas 1998) The preservation of these tracks is thanks to the circumstance that a relatively short time after the early pre-humans (Hominids) walked over an ash layer moistened by rain and produced the tracks, the adjacent volcano Sadiman erupted more ash and filled up the tracks. Additional layers followed and facilitated the preservation. Subsequent geological erosion processes occurred over time so that the first track became exposed in 1978, allowing us to artificially preserve the surroundings and in total more than a hundred human footprints.

What do these footprints tell us? To the right, an upright being walked in short, slightly irregular, rather measured steps (F2), foot 2. To the left, a being with feet of only half the size was walking (F1), foot 1, pretty consistently in step with the larger being. The distance between those two tracks is short enough that the larger being could hold hands with the smaller being, who was probably slightly to the rear. The special thing about the tracks of the third, also more likely smaller being (F3), foot 3, is that he/she walked consequently and successfully inside of the footprints of the first larger being. What did we probably encounter? An image of active imitation as it is typical for the human being (compare to Schad 1985)? The intentional lengthening of the pace or

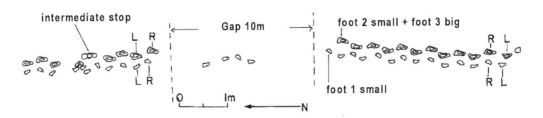

*Fig. 9.19*: The oldest document of the upright stance: the 3.5 MY old footprints from Laetoli, Central Africa. The tracks F2 (foot 2) and F3 (foot 3) are larger and less clearly contoured. It is believed that the slightly smaller individual F3 walked in the footsteps of the individual F2. The clear imprints F1 (foot 1) originated most probably from a child. (according to Agnew and Demas 1998, supplemented).

walking in the footprints of the predecessor was up to now never observed in wild animals. Perhaps we should note that the researchers found the oldest documentation of human education in Laetoli—especially of the activity of imitation shown by the developing child. Using this example, it becomes clear for the students that beings that reached complete uprightness should be called humans.

When discussing the uplift/erection of the human, the following book can be very helpful: *Evolution of Human Walking* (Lovejoy 1988). When comparing the chimpanzee, the hominid find "Lucy" (3.2 MY), and the modern human, "Lucy" ought clearly be classed with the human. This is even true for the mode of birth, as the head of *Australopithecus afarensis* evidently had to slip through the pelvis in a double rotation just as occurs in modern-day human births.

### The Second Step: Hand-Axes as Documents of Human Thinking and Action

Mainstream science regards stone tools or hand-axes almost exclusively as technical equipment. It is thanks to Wolfgang Schad that we begin to see these oldest documents of culture as cult or ritual objects. (Schad 1985, 1996) The oldest rocks hewed by human hand—almost entirely very hard quartz rubble and fine basalt—come from the central rift valley of Africa, embedded within 2.6 to 2.4 MY old sediments. These proto hand-axes are rounded, beveled either in part or all around. While it is still possible to assign a tool character to pieces that are beveled on only one side, this is not possible for pieces that are processed on all sides (see Fig. 9.20). It is therefore more likely that they are an expression of the human design of the sphere, the archetypal form of everything alive.

These artifacts are assigned to *Homo habilis* or *Homo rudolfensis*. The delicate skull form of *Homo habilis* (Fig. 9.21) clearly differs from the robust form of the Australopithecines who was completely upright and we know did work on stones. The modernity of the *Homo habilis* skull is astonishing when compared to the skulls of the younger *Homo erectus* and even *Homo neanderthalensis*. Typical is the retention of all striking bone structures. It gives the impression of a first announcement of modern man, but soon faded away as *Homo habilis* was found only between 2.1 and 1.5 MYA in an obviously weak population.

At the southern end of the Lake of Gennesaret lies an amazing finding site where the typical variety of proto hand-axe forms were found with bone fragments of *Homo habilis* in 2 MY old sediments. The human capable of working with his hands (*Homo habilis*) was obviously the first of our ancestors to leave the African continent and reach Asia minor along the Arava-Jordan rift valley.

When trying yourself to cut ancient or more modern hand-axes by knocking stone on stone, it becomes obvious that this activity trains thinking, imagination, and will. Before the stroke, one needs to have a clear inner image of the target, which means to chip off a certain piece from the rounded rubble (activity of imagination). The strike itself has to be distinct, forceful and exact. This requires a high level of concentration and the conversion of imagination into a willful act. The manufacturing of hand-axes is therefore the ideal exercise to train the soul activities of thinking and will, accompanied by subtle emotion. If we think about the following, we will see that this could happen only when using the hardest of all rocks. At that time, the soul activities thinking, feeling, and

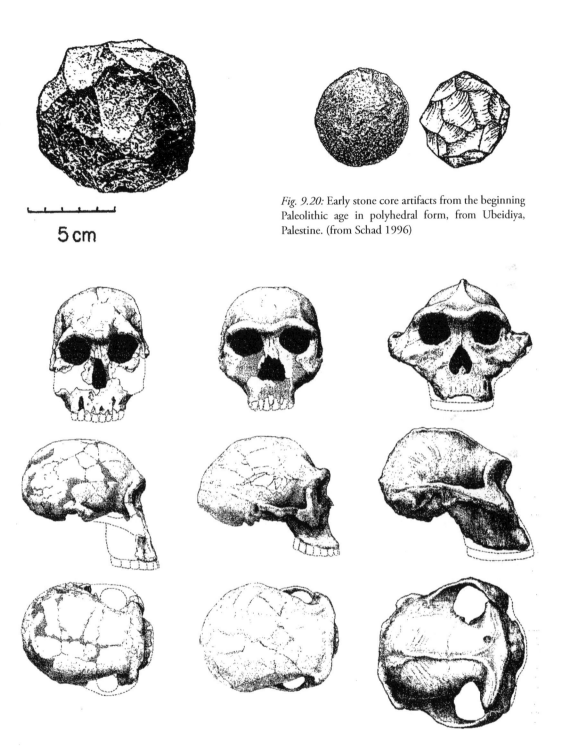

*Fig. 9.20:* Early stone core artifacts from the beginning Paleolithic age in polyhedral form, from Ubeidiya, Palestine. (from Schad 1996)

*Fig. 9.21:* The delicate small skull of *Homo habilis* (left) was found in Koobi Fora, Kenya (KNM-ER 1470) and is approx. 2.1 MY old. Compared to this, the approximately 1.5 MY old skull of *Homo erectus* (middle) looks more archaic despite being much younger. It was excavated in Nigeria (KNM-ER 3733). *Australopithecus robustus* lived at the same time as *Homo habilis* in southern Kenya (right). This skull shows more ape-like characteristics such as a very wide jawbone and a high sagittal crest (KNM-ER 406). The left skull has a brain volume of 780 grams, the middle skull 850 grams, and the right skull a brain volume of 550 grams. (from Walker and Leakey 1978)

wanting could not be experienced inwardly. These processes needed to be externalized. The chipped stones were like extroverted memories of cumulative capabilities.

It is also noticeable that the products are not found at work and residential sites as would be expected for tools, but dispersed over wide open savannah fields. The hand-axes were created, and, after the work was done, they were just left behind. The most important act, the formation of capabilities, is done when the act of creation is concluded. This consideration gains even more plausibility when looking at the next step in the making of hand-axes. The early *Homo erectus* created sophisticated almond-shaped hand-axes of considerable size (Fig. 9.22). Up to 1.4 MY old, these delicately worked objects, with fine edges all around, are harmonic in shape and present the geometrical elements of point (vertex), straight line, and arc. The stone-axe of such design—with sharp edges all around— would be dangerous to use as a tool. This type of hand-axe was produced in a surprisingly uniform style for more than 1 MY in Africa and later in Europe and Asia. It is especially interesting that such well-proportioned hand-axes were found repeatedly as burial objects until fairly recent times.

This clearly points towards a ritualistic act of manufacturing as well as to a ritualistic function of these stone hand-axes. With the beginning of the Middle Paleolithic (250 TY = Thousand Years), a new technique appears as well as a larger variety of hand-axes. First a large pebble was decapitated. Then chips were knocked off with well-aimed strikes on the crest, from which now thinner and smaller hand-axes were made in a further process of touch-ups of the edges, or what is now called reduction technique or stone knapping. This technique was developed by the late *Homo*

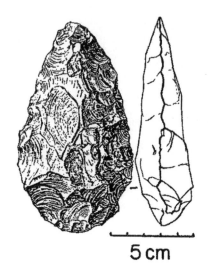

Fig. 9.22: Almond-shaped two-sided classic stone-axe from the middle lower Paleolithic (Acheulean, 0.7 MY) Qumm-Qatafa cave, Palestine. (from Schad 1996)

*erectus* but was also used by the early *Homo neanderthalensis* and *Homo sapiens* (Fig. 9.23). Many of these stone implements with specialized shapes were now really used as tools, which are also found dispersed over large open fields, where also occur the raw materials such as chert nodules, quartzite, fine-grained basalt or rock crystal.

Around 35 TY ago, a sudden increase from approximately 10 to 40 types of stone tools occurred. This marks the beginning of the Upper Paleolithic (refer to Fig. 9.24). The devices were much smaller and are a sign of the unsurpassed crafts art of the *blade technology of hand-axes*.

In summary, the following picture can be drawn. First universal sphericities developed, then the sphere differentiated from an egg form to the classic bifacial hand-axe with an apex on one side and a semicircle on the other side (almond-shape). However, the stone was always worked in such a way that the core was the result (sculpturing). With the reduction technique, the forms differentiated into tools such as scrapers, knifes, sickles, and other blades. However, the classic hand-axe shape was always also produced.

It is a remarkable fact that the unusable core, the waste product, was deposited ritually. This is described by Schad for a finding site in En Avdat in the Negev desert. (Schad 1996)

This is another example through which students realize that the path originates from a universal basic design and diversifies into specialization over time. First comes the universal formation of the capabilities of soul forces, then a specialized know-how arises, which leads to the first forms of technology.

## The Third Step: Development of Language

We can draw conclusions about the ability of our ancestors to articulate from studying the cavities inside the skull, the tooth position, the form of the roof of the mouth, and when possible, the shape of the larynx. One group of researchers theorizes a gradual increase in the powers of articulation from the time of the appearance of *Homo habilis* and *Homo erectus*. Another theory postulates a sudden increase in an intensive language ability from the onset of *Homo presapiens* and *Homo preneanderthalensis*, starting at approx. 200 TY. (Damasio and Damasio 1992, Joseph and Mufwene 2008; Lewin 2005, chapter 7; Tattersall 2000)

It is possible to see in the skull of *Homo habilis* (refer to Fig. 9.21) that *Broca's area* was already present. The question is how was it used. Let us look once again at the development of hand-axes, accompanied by a steady increase in brain volume from 400 g (*Australopithecus*) to 700 g (*Homo habilis*) and from 1000 g (*Homo erectus*) to 1600 g (*Homo neanderthalensis*). Then, from around

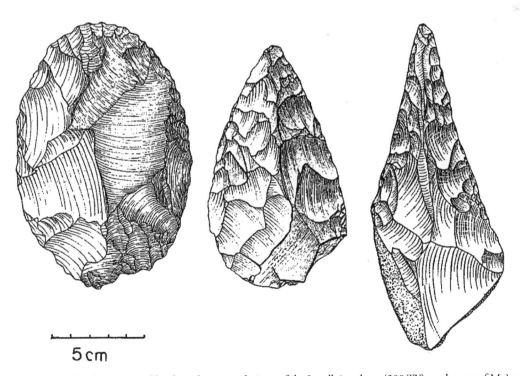

5 cm

*Fig. 9.23:* Hand-axes created by the reduction technique of the Levallois culture (200 TY), settlement of Maàyon Barukh, Jordan Valley. (from Schad 1996)

1.4 MY on, add to the picture of uncertain documents of man's control over fire by *Homo erectus*, and from perhaps 800 TY on, add the first ritual sites.

All these achievements of human activity can be seen in the social association of man to man. When thinking leads to doing within a social community, the use of language can be expected—even if it develops slowly. The fact remains that early mankind documents a three-step in the order of: uplifting—thinking—speaking, as we still observe today in the development of children.

### The Neanderthals: Misunderstood Beings

The research history of *Homo neanderthalensis*, the Neanderthal, is an example of handing down improper ideas. Up to the 1980s, the Neanderthal was presented as a not fully upright, bulky-built and brutally aggressive creature, viewed as somewhat of a degeneration of the human being, in other words as an animal-like, not yet human-like ancestor. It is possible to include a historic excursus at this point to show how ideological ideas can drive exact science into a corner or skew interpretations of research results. Modern archeology describes the Neanderthal as a separate species, found in Europe from approx. 500 TY on as pre-Neanderthal and from approx. 200 TY on as classic Neanderthal in Europe, Asia, and Africa. [Hublin 2009—tr.] Neanderthals were different in the sense that they settled in climatically colder regions at a time when climate and the related landscape fluctuated greatly between severe glacial and very warm interglacials (refer to table A). The achievement of autonomy towards the

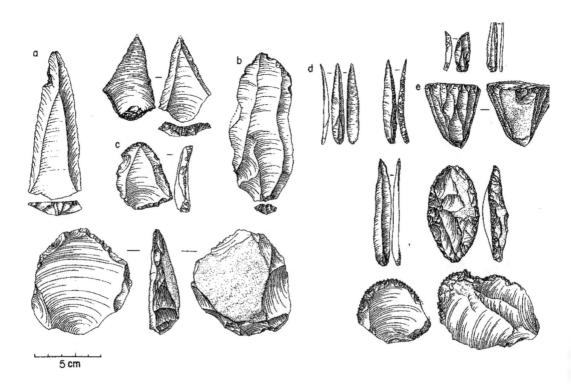

5 cm

*Fig. 9.24:* Small reduction devices with a broad diversity of uses; En Avdat, Palestine. a) hand cones, b) blade, c) scraper, d) scratcher, e) reduction core. (from Schad 1996)

natural climate was one of the outstanding accomplishments of the Neanderthal. To achieve this, the Neanderthal used natural caves, which occur predominantly in limestone areas.

In some of these caves, ritual graves have been found that allow us to draw a picture of a significant social ability. The Neanderthal cave Drachenloch (above Vättis, Switzerland) is at an altitude of 2550 m above sea level. During the second to last alpine glaciations (60 to 50 TYA), this cave was slightly above the ice shield and was inhabited alternatingly by cave bears and Neanderthals. Stone case graves were found in the inner part of the dark cave. In one of these, several bear skulls were found ritually placed in east-west alignment, clearly a sign of worship and not an attempt to brutally eradicate cave bears. Further, the tooth forms of Neanderthal skeletons clearly document that the Neanderthal and his ancestors were mainly vegetarians.

The prologue of the book *Human Evolution* (Lewin 2005) outlines a life picture of the social culture of the Neanderthal based on the grave finds in the Shanidar cave in northern Iraq. Due to the burial objects—floral wreaths and well-proportioned, almond-shaped hand-axes—we can assume that a human being, who, according to his bone configuration, experienced multiple physical challenges, was revered and ritually buried. As a physically challenged being, the "Old Man of Shanidar" was cared for for many years and after his death he was buried with extraordinary efforts. In a grave of a mother and child found in a cave in eastern Turkey, the child was also buried with a garland of wild flowers. The archaic art of the Neanderthal—simple, rhythmically repeating scratch patterns on bones and stones—supports this more recent view of the Neanderthal as a socially highly developed human being.

It is astonishing that over a long period of time (100 to 30 TY), the Neanderthal lived in Palestine together with *Homo sapiens* without intermingling or fighting. In the Mehl cave near Nazareth, both Neanderthal skulls and early forms of modern *Homo sapiens* were found in strata of approximately the same age (Fig. 9.25). The skull form of the Neanderthal is very similar to the skull form of the younger *Homo erectus*. It seems as if the Neanderthal might be an extinct branch of Hominids. Early Neanderthal skulls were more similar to *Homo sapiens* as the later forms shortly before the extinction of the Neanderthal (27 TY).

The following achievements can be attributed to the Neanderthal: 1. Colonization of regions of alternating glaciations and interglacials; 2. Development of a high social culture with ritual burial practices; 3. Colonization of western Europe in the largest population of any hominid so far. (Schrenk, Müller, and Hemm 2009)

### *The Appearance of Art during the Last Ice Age*

As indicated by 110 MY old bone finds in the Qafzeh and Skhul caves in Israel, modern *Homo sapiens* obviously settled first in Palestine. The modernity of the skull is astonishing (refer to Fig. 9.25). Special characteristics are a prominent chin, a well-proportioned arch of the back of the head (occipital), and the lack of large cheek (zygomatic) bones and brow ridges. A *Homo sapiens* skull found as early as 1924 near Singa in Sudan is now dated at 130–150 TY. (Tattersall and Schwartz 2009) *Homo sapiens* barely managed to survive until the beginning of the Aurignacian (35 TY), shortly before the cooling in the most recent glacial period. And all of a sudden, we find documents of art all over the world.

We find small sculptures of animal and female human figures, either carved in bone and some made from limestone. Half reliefs in clay and stone scratch technique developed, and slightly later, colored cave paintings. Necklaces also testify to artistic and creative work, and a piece of hollow long bone with boreholing documents the earliest musical instrument (27 TY). The culture magazine *Du* committed the August 1996 issue to ancient art and justifiably called it *Mastery Out of Nothing.* (Clottes 1996; see also Clottes 2008) [A good online resource is: http://donsmaps.com/indexpaintings.html

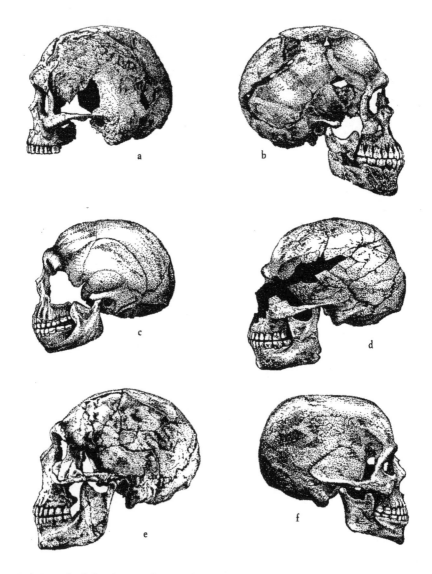

*Fig. 9.25:* Early Neanderthal and sapiens humans from the Palestine region: a) *Homo sapiens* from the Qafzeh cave (110 TY), b) sapient human with Neanderthal characteristics and prominent chin from the Qafzeh cave (approx. 100 TY), c) Neanderthal from the Tabun cave (120 TY), d) Neanderthal from the Skhul cave (approx. 100 TY), e) Neanderthal from the Amud cave (approx 45 TY), f) *Homo sapiens* from the Erq-el-Ahmar cave (approx 25 TY). (from Schad 1996)

—tr.] Cave paintings in southern France and the Pyrenees from the last ice age represent a highlight of art altogether. These congenial line drawings and color pictures of subjects from the ancient animal world are testament to the fact that humans of that time were able to completely feel the animal beings as if they were inside of them. Let us imagine that most of these paintings were done without light, and could therefore not be corrected by distant observation while they were being created. This also explains why many of the most beautiful pictures were painted on top of each other, something that would never enter the mind of modern times artists. The act of painting must have had also cultic and ritual meaning. Paintings and half reliefs are wonderfully integrated into the nature-given cave architecture so that the artists often must have been in uncomfortable positions during the act of painting. Pictures and sculptures were obviously not created to be viewed, or they would appear in more accessible sites.

A motif known from the discussion of the culture of hand-axes reappears: The act of creation by the artist is important—not simply the use of the product. It was all about experiencing the soul of the animal in such a way that it was possible to externalize it. This is another act of the awakening of mankind.

## Evolution and Descent

### From the Universal to the Specialized

In the previous part of this chapter, we passed through the course from early traces of life until present time. By doing this, we discovered over and over again that universal life forms were present initially and that specialized forms appeared later on. In this, the universal entity has the character of perfection, with the form shape curling around the sphere. The effects of self-activity of universalists were essential. One need only to think of cyanobacteria, which enabled modern climate, or of hand-axes, the archetype of the worlds of art and work. More highly differentiated, more complex specializations entered the stage on earth at times when likewise more specific and diverse habitats were available.

In this respect, the entire early world had an universal life character. New eras of time always developed through polarization; the separation of plants and animals, for example, marked the beginning of the Paleozoic. The next polarization separated land and water organisms.

### The Human Being in Balance between Self-Being [Individuality] and World-Being

When describing the simplest organism, we employ the concept of the *shell*. The shell differentiates an outside world from the inside, which is home to the self-activity of the organism. In order to get to life processes, the shell must add the function of permeability from the outside to the inside and vice versa to its function as a boundary between inside and outside. It is this selective exchange that constitutes metabolism. Part of the world is taken in, is destroyed, and will be used to build up the substance of the inside life. Waste products of these processes have to be given back to the outside world by passing out through the shell. Hence, the outside world will change also.

The biological species *prokaryotes* wanted to realize this principle, and had the opportunity to do so during the first rock formations. Separation of world-being from self-being as an inward gesture was not very pronounced yet. At the transition between Proterozoic and Paleozoic, creatures appeared that were either flowing out into the world or wanted to be more turned away from the world. The separation of animals and plants became possible. By invagination of the differentiating shell, animals created interior spaces where they can live their self-being parallel to their self-activity. The soul, the world of emotions and feelings, received a physical residence. Plants on the other side, developed the leaf principle, turning the differentiating shell outwards. This enabled plants to reach far into the world with their metabolism and self-being. World-being became possible.

It is characteristic of both plants and animals that they can predominantly but never entirely live out world-being or self-being. This is so because extreme one-sidedness would break up the metabolism and cause death. The breaking-up of life's metabolism or the incidence of death now leads to the world of minerals.

If we now look at the human being, we realize that he/she has the ability to balance self-being and world-being through spiritual activity. This also opens up the possibility of actively being *eternal*. Through the sense organs the world reaches inside the human being. Man overcomes the initial incomprehension [separation] of/from the world by means of thinking, and reaches an active reconnection with the world in the act of knowledge. The experience of the *balance between self-being and world-being* is reached through the necessary dealing with antipathy (standing opposite of the world) and sympathy (connecting to the world). The experience of being eternal develops by penetrating the intrinsic nature of the world in a spiritual, self-active way. Truths that are achieved by insight into the nature of things have the character of the eternal (comp. Fig. 9.26). In this way we can see that the *human being is the union and origin of all realms of*

*nature*. In the geological course of life, we get acquainted with one-sidednesses that fell to early off from the ideally-thought spiritual-soulful human being.

The world phenomena on earth that vary during the course of times are the effects of the developing human from a spiritual sphere to a spiritual-soulful sphere, and further to the spiritual-soulful alive, and finally into a state in which the human unites with the earthly world of matter. A qualitatively different presentation and further thoughts can be found in *An Outline of Esoteric Science*, especially in the chapter "Man and the Evolution of the World." (Steiner 1910a, Steiner and Creeger 1997)

### *Motifs of Evolution*

Besides the differentiation from the uniform universal to the specific, there are additional motifs that can be read from evolution. When observing the way of life of animals from the explosive diversity in the Cambrian until present time, the following becomes apparent. In the course of time towards the present, species occur step by step that express an *emancipation from the natural environment*: active movement, skin armor, body heat, individual embryonal space. The same but in reduced form can be found in plants in the shaping of organs and metabolic processes, which make plants slightly less dependent on the environmental climate.

We find the same sequence when looking at the fossil history of man. The early African hominids were still completely included into the environment of the rift valley climate. *Homo erectus* and the Neanderthal were increasingly emancipated from the climate. By looking at the physical-geographical and geological-earthly life conditions at the time of biological novelties, we can experience how the physical earth—subject to ongoing transformations due to life processes—gave off supporting and inhibiting conditions for the new unfolding of life.

The world as it was had no causal effect on the development of life, but provided the conditions for the unfolding of the species, which was already ideally pre-structured. We know this phenomenon well in the human practices of agriculture and animal keeping. We need to establish the conditions under

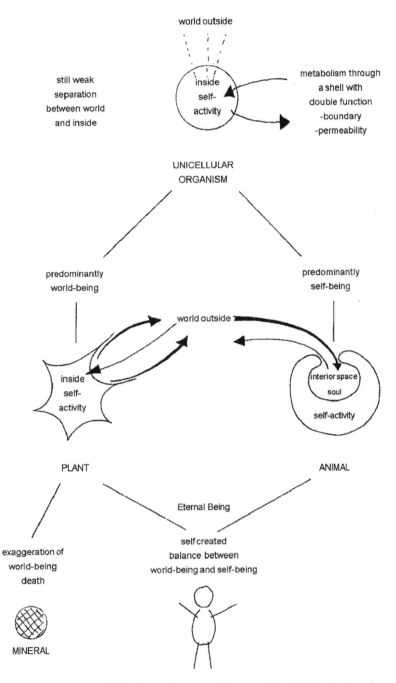

*Fig. 9.26:* Schematic representation of the relation between self-being and world-being in the realms of nature and in the human being.

which the species can develop best. We can read this from the interrelationship between the physical earth and the unfolding of life and can learn from this for the future. In this way, a basic understanding for the development of the world by and with the human results from the discovery of the geological sequence.

### The Human Being between Procreation and Death Compared to Biological Evolution

If there is an opportunity to discuss or review the embryonic development of the human being with the class, this would correspond very well at this point. We find the geological steps from unicellular organisms to multicellular water creatures and further to aquatic animals and creatures with aerial lungs also in human embryonic development until birth. The three-step of walking—thinking—speaking are steps of an artistic activity, and the steps of the creation of consciousness of early man find their expression in the development of the infant to childhood and adolescent. The students will discover in astonishment that humanity arrived in the 20th century at approximately where they are standing at the end of their schooldays.

Cave on the Yangtze River, China. Photo by DSM

# Appendix:

## Tables A – E

Tabular summary of geological, climatic, cultural and biological events, embedded in a linear and logarithmic scale.

Column 1 sets the linear time scale in thousand years (TY; table A) and million years (MY; tables B to E). In tables A and B, column 2 indicates the alternation between glacial and interglacial periods. Tables C to E show geological events (left) as well as climatic events (right). MB indicates an enhanced formation of seafloor rocks at mid-ocean ridges.

In columns 3 and 4, times are given in logarithmic form. The left column (3) shows the number of the Platonic [Great] Year (25,920 years) as multiplications over 3x7 steps, beginning at the end of the ice age 12,000 years ago. The numbers in the right column (4) arise, when starting in table E and dividing the number of 4600 million years MY through a 3x7 sequence of steps, each divided by 2, until you reach 35,156 in table A.

In tables A–C, the events column sdocument the geological age of hominids (left), specific site of discovery (middle) and cultural achievements (right).

In tables D to E, the events columns are distinguished into notes about life crises (left), atmospheric conditions (middle) and outstanding life events (right).

Surprisingly, events can be assigned to numbers of either the multiplication- or division column. This means that the sequence of events correlates with the logarithmic time scale.

## Abbreviations:

MY = million years

TY = thousand years

MB = maximum in the formation of seafloor rocks

A. = *Australopithecus*

H.s. = *Homo sapiens*

H.n. = *Homo neanderthalensis*

## Table A

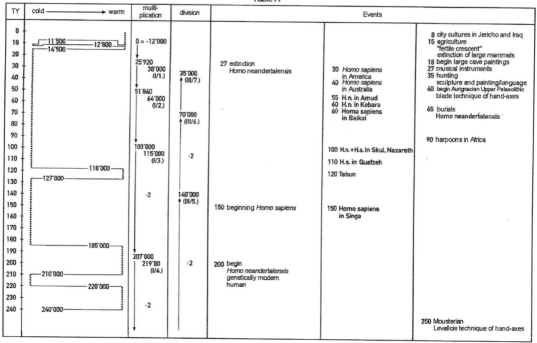

| TY | cold ⟶ warm | multi-plication | division | Events |
|---|---|---|---|---|
| 0 | | | | 8 city cultures in Jericho and Iraq |
| 10 | 11'500 — 12'800 | 0 = -12'000 | | 15 agriculture "fertile crescent" extinction of large mammals |
| 20 | 14'500 | | | 18 begin large cave paintings |
| 30 | | 25'920 | 35'000 ↑(III/7.) | 27 extinction Homo neandertalensis — 30 Homo sapiens in America — 27 musical instruments |
| 40 | | 38'000 (I/1.) | | 40 Homo sapiens in Australia — 35 hunting sculpture and painting/language |
| 50 | | 51'840 | | 40 begin Aurignacian Upper Palaeolithic blade technique of hand-axes |
| 60 | | 64'000 (I/2.) | | 55 H.n. in Amud — 60 H.n. in Kebara — 60 Homo sapiens in Baikal |
| 70 | | | 70'000 ↑(III/6.) | 65 burials Homo neandertalensis |
| 80 | | | | |
| 90 | | | | 90 harpoons in Africa |
| 100 | | 103'000 | | 100 H.n.+H.s. in Skul, Nazareth |
| 110 | | 115'000 (I/3.) | ·2 | 110 H.s. in Quafzeh |
| 120 | 118'000 | | | 120 Tabun |
| 130 | 127'000 | | | |
| 140 | | | 140'000 ↑(III/5.) | |
| 150 | | | | 150 beginning Homo sapiens — 150 Homo sapiens in Singa |
| 160 | | | | |
| 170 | | | | |
| 180 | | | | |
| 185 | 185'000 | | | |
| 190 | | 207'000 | | |
| 200 | | 219'00 (I/4.) | ·2 | 200 begin Homo neandertalensis genetically modern human |
| 210 | 210'000 | | | |
| 220 | 220'000 | | | |
| 230 | | | | |
| 240 | 240'000 | | ·2 | |
| 250 | | | | 250 Mousterian Levallois technique of hand-axes |

## Table B

| MY | cold ⟶ warm | multi-plication | division | events | | |
|---|---|---|---|---|---|---|
| 0.0 | 0.01 | | 0.140 | | | 0.25 begin Mousterian Middle Paleolithic Levallois technique |
| 0.1 | WÜRM = WISCONSIN 0.12 | | | (III/5.) | | 0.30 low larynx first stone scratches? first proto-sculpture? |
| 0.2 | 0.127 RISS II = SANGAMONIAN 0.185 | 0.207 | 0.281 | | | |
| 0.3 | 0.24 RISS I = ILLINOIAN 0.3 | 0.219 | (III/4.) | 0.3 End Homo erectus | 0.30 Petralona | |
| 0.4 | 0.335 MINDEL II = YARMOU 0.39 | 0.415 | | | | |
| 0.5 | 0.42 MINDEL I = KANSAN 0.46 | 0.427 (I/5.) | 0.563 | 0.5 begin Homo pre-sapiens Homo heidelbergensis | | |
| 0.6 | | | (III/3.) | | | |
| 0.7 | 0.75 | ·2 | | | | 0.70 many camp fire sites |
| 0.8 | GÜNZ = NEBRASKAN | 0.829 | | | 0.75 Peking, Homo erectus / 0.80 Homo erectus in Europe | |
| 0.9 | 0.9 | 0.841 (I/6.) | ·2 | | | 0.90 more than 10 kinds of hand-axes |
| 1.0 | | | | 1.0 End Australopithecus robustus | 1.00 Homo erectus in China and Java | 1.00 begin Acheulean Lower Palaeolithic |
| 1.1 | | | 1.125 | | | |
| 1.2 | | | (III/2.) | 1.2 End Australopithecus boisei | | |
| 1.3 | | ·2 | | | | 1.40 first sporadic camp fire sites |
| 1.4 | 1.45 | | | | | 1.50 begin hand-axe diversification |
| 1.5 | DONAU - NEBRASKAN? | | | 1.5 End Homo habilis | | |
| 1.6 | 1.6 | 1.66 | | | 1.60 Turkana Boy Homo erectus | |
| 1.7 | | (I/.7) | ·2 | | | 1.80 Broca's speach center |
| 1.8 | | | | 1.8 begin Homo erectus end Homo rudolfensis | | |
| 1.9 | 1.9 | | | | | |
| 2.0 | | | ·2 | 2.0 begin Australopithecus robustus end Australopithecus africanus | | |
| 2.1 | BIBER | | | | | |
| 2.2 | | | | | | |
| 2.3 | 2.3 | | 2.250 (III/1.) | 2.3 begin Homo habilis / 2.3 end Homo aethiopicus | | |
| 2.4 | | | | 2.4 begin Australopithecus boisei | | 2.50 first archaic hand-axes proto-hand-axes |
| 2.5 | | | | | | |

188

## Table C

| MY | Geology | multi-plication | division | events | |
|---|---|---|---|---|---|
| 0 | | | | | |
| 1 | | | | | |
| 2 | begin ice age | 1.66 (I/7.) | 1,125 (III/2.) 2,250 (III/1.) | 2,0 end A. africanus<br>2,3 begin Homo habilis?<br>2,4 begin A. boisei<br>2,5 begin A. garhi<br>2,5 begin Homo rudolfensis<br>2,8 end A. afarensis<br>3,0 begin A. africanus | 2,5 first proto-hand-axes |
| 3 | renewed uplift of the Alps | 3,32 (II/1.) | | 3,9 end A. anamensis<br>4,2 begin A. anamensis<br>4,4 Ardipithecus ramidus? | 3,5 footprints from Laetoli |
| 4 | | | 4,500 (II/7.) | | 4,0 uplifting / erection?<br>maybe much earlier?<br>see S. tschadensis |
| 5 | glaciation of high mountains | ·2 | | | |
| 6 | | 6,64 (II/2.) | | | |
| 7 | fold of the Jura | | :2 | ~7 Sahelanthropus tchadensis<br>separation between<br>African apes<br>and hominids? | |
| 8 | | | | | |
| 9 | | ·2 | 9,000 (II/6.) | | |
| 10 | | | | | |
| 11 | strong cooling of the Alps | | :2 | | |
| 12 | | | | | |
| | | 13,3 | | | |

## Table D

| MY | Geology | multi-plication | division | events | |
|---|---|---|---|---|---|
| 0 | ice age | | 9,0 (II/6.) | | 4 uplifting / erection - genus Homo |
| 10 | Alps become high mountains | 13,3 (II/3.) | 18,0 (II/5.) | | 7 seperation between<br>African apes<br>and hominids |
| 20 | North Atlantic breaks open | 26,5 (II/4.) | | | |
| 30 | | | 36,0 (II/4.) | | 30 primates climb on trees<br>large mammals<br>sudden diversification<br>of mammals<br>Placentalia<br>extinction of dinosaurs |
| 40 | | 53,0 (II/5.) | 53 end life crisis | | |
| 50 | | | | | |
| 60 | meteorite event | | 65 maximum life crisis | | |
| 70 | MB sinking of sealevel | ·2 | 72,0 (II/3.) 72 begin life crisis | | |
| 80 | maximum oceans | | | | |
| 90 | begin of the alpine fold | | | | |
| 100 | | 106,0 (II/6.) | | | |
| 110 | | | :2 | | |
| 120 | | | | | |
| 130 | | | | | 130 first magnolia-like<br>flowering plants |
| 140 | temperature maximum 23 °C | | 144,0 (II/2.) 140 life crisis | | |
| 150 | | ·2 | | | 150 real birds<br>small mammals<br>warm blooded animals |
| 160 | | | | | |
| 170 | South Atlantic breaks open | | | | |
| 180 | begin E-West Tethys Sea | | | | |
| 190 | Raising of the sealevel | | :2 190 life crisis | | 200 modern bony fish<br>new large reptiles |
| 200 | meteorite event | | | | |
| 210 | PANGAEA | 212,0 (II/7.) | | | |
| 220 | | | | | |
| 230 | MB sinking of the sealevel | warm | | | |
| 240 | | | 287,0 | | |

| | MY | Geology | multi-plication | division | events | | |
|---|---|---|---|---|---|---|---|
| Meso-zoic | 0 | | | ↑ | | | |
| | 100 | 60 orogenesis | | 106 | | | |
| | | | warm | | 144 | 210 end life crisis | current oxygen | 200 begin Gymnosperms |
| | 200 | 190 orogenesis | cool | 212 | ↑ (II/2.) | 225 maxim. life crisis | content reached | 300 ancient conifers |
| Paleo-zoic | | 220 PANGAEA | warm | | 287 | 270 begin life crisis | | 330 reptiles/psylophytes/proto-ferns in Europe |
| | 300 | 220 orogenesis | ice age | | ↑ (II/1.) | 365 life crisis | | 350 carboniferous forests |
| | 400 | | warm | | | | 430 ozone layer | 420 coelacanths. four limbs |
| | | 440 orogenesis | ice age | 424 | 575 | 440 life crisis | | 440 amphibia/life on land |
| | 500 | 570 Caledonian orogenesis | ice age | (III/1.) | ↑ (I/7.) | 500 life crisis | | many land plants |
| Protero- (Vendian) zoic III | 600 | ··· MB | ice age | | | 600 life crisis | 570 7 % oxygen | from 500 first land plants in Siberia |
| | 700 | | | | | | in atmosphere | from 540 rapid diversification of animal phyla |
| | 800 | 750 PANGAEA | | | | | | 630 end Ediacara |
| | | | | 849 | :2 | | | 670 bloom Ediacara |
| | 900 | | | (III/2.) | | | | 800 complex space shapes |
| | 1000 | ··· MB | ice age | | | | | 810 begin Ediacara creatures |
| Proterozoic II | 1100 | 1100 PANGAEA | | | 1150 | | 1100 1 % oxygen | 1100 multicellular organisms |
| | 1200 | | | | ↑ (I/6.) | | in atmosphere | (spheres and tetrads) |
| | 1300 | | | :2 | | | | life on the earth's surface |
| | 1400 | | | | | | | |
| | 1500 | | | | | | | |
| | 1600 | 1650 PANGAEA | | | :2 | | | 1600 fungi |
| Proterozoic I | 1700 | | | 1694 | | | | 1600 female and male cell |
| | 1800 | | | (III/3.) | | | | cells with nucleus: |
| | 1900 | | | | | | | eukaryotes |
| | 2000 | ··· MB | | :2 | | | 2000 oxygen | 2000 oxygen-tolerant |
| | | 2100 PANGAEA | | | 2300 | | in atmosphere | unicellular organisms |
| | | ··· MB | ice age | | ↑ (I/5.) | | red sediments | 2800 proto-algae (filaments) |
| Archean | 3000 | 2600 PANGAEA | | | | | | |
| | | | | 3394 | :2 | | | 3800 stromatoliths |
| | | | | (III/4.) | | | | 4000 photosynthesis - cyanobacteria |
| | 4000 | | | | | | 4000 oxygen release | before 4000 prokaryotes: archaea, |
| | | | | | | | into water | sulfur bacteria? |
| | | | | 27 000 | 4600 | | | 4200 first crystalline rocks and their |
| | | | | (III/7.) | (I/4.) | | | transformation products/carbonatic |
| | | | | | | | | and sulfate sediments |

Canyonlands, Utah. Photo by DSM

# Chapter 10

# Aspects of Global Economy—Grade 12

## Presentation of the Problem

While introducing the paleontology main lesson (chapter 9), we explained why it makes sense to teach the topic of modern global economy in grade 12. After the paleontology main lesson has enabled the adolescents to relate to the earth's and the human's processes of becoming, they are better equipped to perform solid judgments about economic issues. Especially toward the end of their schooldays, it is important to stimulate the young people to perform their *own* judgments from their insights into the nature of the topic. Today's economic life—the financial markets, monetary policy, technological restructuring of jobs, continuously changing currency reforms and, last but not least, the strong entanglement of the economy and world politics—has become so incalculable that even experts cannot describe these factors anymore. Teachers are facing the danger of portraying the underdeveloped material in a spectacular and incoherent way, and thus lead students away from the option to form their own judgments.

Due to this situation, the attempt was made to work on elementary aspects of global nutrition in such a way that still allows students a certain degree of self-processing. This class was taught as a crafts block for a couple of weeks, each with three double periods per week. Good sources are the books by Strahm: *Why Are They That Poor?* (1995) and *Economy Book Switzerland* (1992), as well as *The Maximum Living Wage* (Sax et al. 1997). Strahm's books may contain old numbers that should be updated; however, their didactic design and carrying out are exemplary. [There are, however, two newer editions of what seems to be a mix of both books: *Why We Are That Rich—Economy Handbook Switzerland* (Strahm 2008, 2010) A brief review of the literature available in English indicates that the following resources might be helpful in a similar way: Harrison and ebrary Inc. 2007; Hillebrand 2008; Hoekman and Olarreaga 2007; Isbister 2006; Knutsen 2010; Kohl and Organisation for Economic Cooperation and Development, Development Centre 2003; Lines 2008; Little 2003; Thirlwall and Pacheco-López 2008; Wiggins 2003—tr.].

## World Nutrition

The students have already discussed the topic of industrialization with the examples of textile technologies and the economy of energy as described in chapters 6 and 8. Let us think back to the human science block of the 7th grade (chapter 1). The relationship between the human being and the world was discussed using the explicit key words:

nutrition [food], clothing, sanitation and housing. In the analogy of these topics in high school teaching, the topic of nutrition is still missing.

Proper nourishment of an individual person shall not be the topic of this discussion but rather the question of fairness in the production and global distribution of food. What is a fair price? How can we avoid widespread famine? These are the questions that need to be discussed.

### How Is a Fair Price Made?

Two examples may be thought through in order to approach this question. The relationship between industrial countries and developing countries can be illustrated by comparing the production of wheat and of cocoa.

### Wheat Production

The principle wheat-growing areas are located in the temperate zone of the Northern hemisphere, i.e., in the USA, Canada, Europe, and the countries of the former Soviet Union. These regions are at the same time industrialized and mainly prosperous. Wheat production is highly technological and is based on extreme mono cultures. In North America, the consequences of soil overexploitation became already visible during the 1930s. Wide areas became unfertile by inappropriate soil tillage and dust storms. Continuous gains in yield were possible only with extensive use of fertilizers and seed breeding.

Students know these problems from their gardening classes and the farm trip in grade 11. They are familiar with the methods of organic and biodynamic agriculture, whereby soil enrichment is a priority rather than mass production tolerating the exploitation of the ground.

When we arrive at this topic, inevitably the question of price appears. Students clearly prefer organic or, even better, biodynamic agriculture. When calculating the price for 1 kg of organic wheat, we figure a price that is at least twice as high as wheat produced in industrialized agriculture, because of the amount of labor involved. If we add surcharges originating from international trading and customs, the difference between the price of cheap imports that were transported over vast distances and regionally distributed organic wheat becomes slightly reduced (comp. chapter 10).

If we consider not only the current production but also its sustainability in the future, this leads to the insight that a fair price results from a mode of cultivation that increases natural soil fertility. Today's unfair wheat price is based on an overexploitation of soils and a misguided seed breeding. This is because the monopolization of the breeding of seeds by only two worldwide leading agrobusiness firms leads inevitably to dependency. Almost one fourth of the world's 8.7 billion hectare of agricultural land deteriorated massively between World War II and 1997 (according to The United Nations Environment Program 1997; helpful information and data on their website: http://www.unep.org and in the annual yearbooks available as free PDF downloads; printed copy available for a small fee).

After the students understand that a different way of wheat cultivation will lead to higher prices, the question of affordability arises. Here we could revisit the history curriculum in grade 9. During that block we discussed human rights. Basic human rights include the right to work and the rights to eat and have a place to live. Wheat is produced for all of mankind. For their work, people in the Third World, who are living in areas that are not growing wheat, should receive more than a minimal wage imposed by the industrialized countries. Instead, their work should be honored according to the principle of the *supply of needs*. Their needs would then be calculated as the *fair prices* of food items, and by this the expensive imported wheat would be affordable.

This train of thought raises the relationship between *labor* and *wages*. The enforcement of a worldwide *division of labor has* resulted in the condition that a person is not working for him- or herself but for others. As a specialist, we produce goods for others and not for our own requirements. However, this job can be performed only if other people that take advantage of these goods provide the specialist with the basics for their lives. Such a practice of reciprocal brother- and sisterhood extends far beyond the current notion that we work for appropriate wages. We work for other people and they in return take care that we have all that we need.

Steiner's essay "The Science of the Spirit and the Social Question" contains a clear wording of this main social law:

Now, the main social law set forth by the science of spirit, is the following: "The well-being of a total community of human beings working together becomes greater the less the individual demands the products of his achievements for himself, that is, the more of these products he passes on to his fellow workers and the more his own needs are not satisfied out of his own achievements, but out of the achievements of others. All the conditions within a total community of people which contradict this law must sooner or later produce misery and distress somewhere.

This law holds good for social life with absolute necessity and without any exceptions, just as a natural law holds good for a particular sphere of natural processes. But it should not be thought that it is sufficient for this law to be held as a universal moral law, or that it should be translated into the attitude that everyone should work in the service of his fellow men. No, in actual fact the law will be able to exist as it should only if a total community of people succeeds in creating conditions where no one ever can claim the fruits of his own work for himself, but where, if at all possible, these go entirely to the benefit of the community. And he in turn must be maintained by means of the work of his fellow human beings. The important thing is to see that working for one's fellow human beings and aiming at a particular income are two quite separate things. (Steiner 1905a, translated as: "Anthroposophy and the Social Question," first published in 1982 by Mercury Press, also as "Spiritual Science and the Social Question," published 1958 by Anthroposophic Press, also available in "The Nature of Anthroposophy," published by Rudolf Steiner Publications 1964; both available online at http://rsarchive.org ).

In the last lectures in the book *World Economy*, Steiner addresses explicitly the relationship between labor on agricultural soils and the value of money. He demands that the value of a money bill should be measured with the measure for the performance of a certain work on a defined soil area destined to produce wheat. The value of money could be compared to real soil product and not, e.g., to gold. The gold-backing of money is an unreal process that has no relationship to reality. Currency has to be connected to usable means of production, for which physical labor is performed. The only healthy currency would be the sum of usable means of production.

> We shall find that our currency, representing, as it were, the day-to-day book-keeping of world-economy, will have to be inscribed, let us say: "Wheat producible over a given number of acres," and this will then be equated to other things. The different products of the soil are the easiest things to equate. So you see where it is we must start from—our figures must mean something. It simply leads away from reality if money has inscribed on it: "So much gold." It leads towards reality if it has inscribed on it: "This represents so much Labor upon such and such a product of Nature." For we shall then have this result: Say there is written on the money "x wheat," all money will be stamped "x of wheat, y of wheat, z of wheat." The real origin of the whole economic life will then be made evident. Our currency will be referred to the usable means of production upon which bodily work is done—the means of production of the given economic region. This is the only sound basis of currency— the sum-total of the usable means of production. (Steiner 1922b, 14th lecture on 8/6/1922, translated by Owen Barfield)

There are completely different opinions today about the backing of currency. As the resulting social and economic problems are that significant, new forms of backing are under consideration. The difference between rich and poor has increased continuously, both globally and nationally. The situation has developed into the unreal. As an example, in 1998, the sum of non-taxable capital gains in Switzerland was larger than the sum of all taxable earnings. This means that the few owners of capital became richer without performing labor than all working people in Switzerland together, who received wages and fees for their living expenses! The Swiss National Bank will earn 20 billion Swiss Francs from the sale of one third of their gold reserves. It is obvious that backing with gold has become less important. Without much care to backing at all, Russia throws a large amount of freshly printed rubles into the domestic financial market.

### Cocoa Production

As another example, we will take a look at the economy of cocoa. Most students will have a more direct consumer relation with cocoa than to coffee, which could be another good example. Most cocoa is produced by a few Third World countries that are climatically well suitable due to their location in the subtropical climate zones (Fig. 10.1).

What contributes to the price of 100 g chocolate, produced by a Swiss company? From a typical sale price of 1.60 Swiss Francs [1 CHF ≈ $1.5] only 0.07 CHF is paid to the cocoa farmer. The chocolate company buys the cocoa for about 0.15 CHF in the developing country. The doubling of the price occurrs due to customs fees. The chocolate company sells the 100 g chocolate bar for about 0.8 CHF to the wholesaler. Wholesale and retail trades earn another 0.8 CHF, bringing the consumer- or end price to 1.6 CHF.

When students see these numbers, they often independently come up with the idea to pay 0.15 CHF more per bar. This way we could pay the producing farmer three times as much for his/her bread. Now we can introduce institutions that promote a development policy that makes sense, e.g., the Swiss Max Havelaar-Stiftung or the US TransFair USA, and Equal Exchange. These Third World trade organizations sell coffee and cocoa at approx. 10% higher price, which surcharge goes entirely into the country of production. The money is not just given to the producers, but to the cocoa farmers associated into cooperatives and who receive advice about how to improve their cultivation methods, storage, and marketing. This is an attempt to improve cultivation according to ecological and organic aspects in order to avoid the leaching out of soils and to avoid expensive additives used in more traditional agrobusiness. The motto of consulting is help people to help themselves. Another supportive activity of the Havelaar-Stiftung is the eliminating of speculative transactions which result in price increases. Cocoa that was bought from the cooperatives by Havelaar is not traded at raw materials futures exchanges. (See yearly reports of the Max Havelaar-Stiftung, http://www.maxhavelaar.ch, in German only.)

## How Does Speculation on the Future Exchanges of Raw Products Work?

Raw materials [commodities] can be bought or sold at different times from the particular harvesting times at special commodity exchanges. (Strahm 1995) Traders use price fluctuations that may amount to more than 100 percent either ways (Fig. 10.1). If an increase in prices is expected, the venture capitalist buys at a certain date, e.g., in the spring for the harvest

in autumn. He will sell the goods, however, before the supply is due and therefore make a profit. Buying means that the venture capitalist needs to make an advance payment of 10 percent of the sale price. By investing relatively little money, large gains can be made. If prices are high, and a dramatic drop in prices is expected, the venture capitalist sells contracts for goods at a fixed time. This means that he promises to supply the raw material, which he does not yet have, at an agreed-upon time. Before the delivery deadline, he buys the goods at a low rate and thus when he sells, he makes a profit. The venture capitalist wins both in case of a price increase as well as in case of a dramatic drop in price. Due to this kind of gambling, prices themselves get into larger fluctuation ranges. We get an idea of how big this speculation business is when we realize that in 1982/83 the ownership of the entire cocoa harvest of the world switched 16 times in speculations. This business must be profitable as there are more than 300,000 registered traders for raw materials in US exchanges alone! The ones that suffer from all this trading are the raw materials-producing countries, as they suffer from large price fluctuations. Producer prices also sink due to the higher end-user prices caused by speculation gains.

Sensible development policy would therefore include direct supply contracts between poor production countries and rich consumer countries paying fair prices, thus avoiding commodity futures (compare to Max Havelaar-Stiftung).

## Hunger Is a Problem of Food Dispersal and Poverty

After looking at the production of a basic food and a luxury food, the question about the distribution of produced goods arises. Due to newspaper and TV reports,

| country | cocoa harvest 1989 in 1000 tons | mean life expectancy in years | foreign debt in % of the gross national income | illiteracy rate in % of total population | inhabitants per physician |
|---|---|---|---|---|---|
| Ivory Coast | 700 | 54 | 135 | 46 | 15 000 |
| Ghana | 300 | 56 | 47 | 40 | 10 000 |
| Nigeria | 160 | 52 | 102 | 49 | 16 000 |
| Cameroon | 120 | 56 | 45 | 44 | 16 000 |
| Brazil | 390 | 65 | 30 | 22 | 17 000 |
| Malaysia | 250 | 70 | 56 | 21 | – |
| Indonesia | 120 | 58 | 62 | 23 | – |
| reference country Switzerland | 0 | 77 | 0 | 1 | 400 |

| | 1982 | 1984 | 1986 | 1988 | 1990 | 1992 | 1994 |
|---|---|---|---|---|---|---|---|
| cocoa price in US Dollar per ton | 200 | 2450 | 2100 | 1750 | 1250 | 1200 | 1350 |
| cocoa production in 1000 tons | 1700 | 1500 | 2000 | 2200 | 2400 | 350 | 2500 |
| world resources in 1000 tons | 750 | 500 | 700 | 900 | 1500 | 1600 | 1400 |

Fig. 10.1: Characteristics of cocoa production. The upper table shows the clear discrepancy between cocoa producers and a chocolate-producing and consuming country (Switzerland). The lower table shows that cocoa prices fell by one third within a decade. The dramatic drop in prices was the stronger the greater the world cocoa reserves (collected by students using the Fischer Weltalmanach 1985–1996).

most students know about famines in many parts of the world. Is there not enough soil for the production of food? World Bank and FAO (Food and Agriculture Organization of the UN) disagree (Sax et al. 1997; de Ridder et al. 2004; Lal 2008) [Newer and English references in this chapter added by the translator.] Up to very recently, food production kept up with population growth. Since 1995, however, grain production has stagnated which has led to an increase of export prices up to 50 percent. In contrast to this, export prices for products from the Third World fell in the majority of cases (Fig. 10.1). This increased the gap between rich and poor even more.

In the countries of Latin America, South Asia, East Asia except China, the Middle East, and Africa south of the Sahara, only 25 percent of the potential usable land area is used for agriculture. There would be a potential to increase production if water as a prerequisite for agriculture were more available. In the regions that are suitable for agriculture, i.e., that are the temperate latitudes of the rich industrialized nations on the Northern hemisphere, the area of arable land is reduced in context with industrialization and an over-exaggerated implementation of agricultural techniques. Experts in water management stress the necessity of a dramatic restructuring in the near future (Liu, Zehnder, and Yang 2009; Zehnder, Yang, and Schertenleib 2003): For climatic reasons, modern industrial nations should become once again predominantly agricultural nations. Agriculture should be conducted only where no artificial irrigation is necessary. This is because groundwater will be the first resource on earth that will become scarce within the next few years. Clean water will become so valuable that it may not be used for artificial irrigation anymore. On the other hand, industrialization should be moved to climatically drier and sunnier areas. This would also provide the significant advantage that much of the industrial need for energy could be readily provided by efficient solar energy devices built right within industrialized areas. However, the demand for such structural changes is simply ignored, as it is in diametric opposition to the economic and agricultural policies of the industrialized countries, e.g., the European Union.

What conclusions can be drawn from the interpretative outline so far? In order to guarantee nutrition for all of humanity, agriculture will have to be conducted in a sustainable and ecologically justifiable way in areas where sufficient precipitation occurs to support the cultivation of the specific products. Regional and national interests will have to give way to global solidarity and cooperation. For the time being, poverty needs to be fought through debt relief and restructuring of custom fees.

This shows that the principle of brotherhood in a cooperatively organized economy, as demanded by Steiner in context with his research of the threefold, could currently develop spontaneously through the understanding of the crisis of civilization. In the above mentioned *World Economy*, Steiner assumed two basic prerequisites: on the one side, the number of people on earth, and on the other side, the soil area suitable for agricultural use. Every human being should be eligible for as much basic food items [such as wheat] as can be produced according to the soil area that results from the ratio between the total population and the agricultural soil area. The value of money should be measured in labor performance per agricultural area. Under these circumstances, justice would prevail if, in addition, products would be allocated in a fair way. At the turn of the millennium, approx. 9 billion hectares of land are available for approx. 6 billion people, or 1.5 hectares of soil per inhabitant. On this land either wheat for eight people, milk for two people, vegetables for 10 people, or meat for two people can be produced. (Sax et al. 1997; Kukathas 2009; Beste 1997; Loske and Bund für Umwelt und Naturschutz Deutschland 1996; Sachs, Bund für Umwelt und Naturschutz Deutschland and Wuppertal-Institut für Klima Umwelt Energie 2009; Kracht 1999; Pimentel and Pimentel 2008)

### Grains as Feed for the Livestock of the Rich

Many students will be dissatisfied with the discussions so far. Being impatient, they may demand solutions that could be applied for faster results than what would be possible in restructuring the world economy. In addition, they may want to add aspects to the discussion that allow them to be active in their own external life, such as supporting the Havelaar Foundation.

When following the path that globally-grown grain take, we find some astonishing figures. (Haldimann, Dietrich, and Erklärung von Bern, Zürich 1992; Strahm 1992b; Kukathas 2009; Tansey and D'Silva 1999; Rifkin 1992) In 1981, 47 percent of the global grain production was used as animal feed. These 590 million tons of feed grain could have fed 2.5 billion people, or approx. the number of people living in all the developing countries other than China. If, for example, these quantities of grain were not used as animal feed, each person in North America could feed two people elsewhere, and each Swiss could feed one other person somewhere on the planet.

This situation becomes even more drastic when we calculate how much animal feed the European Union imports from malnutritioned developing countries. About half of all soy, oilseeds, oilcakes, and grain substitutes such as tapioca, corn gluten and molasses feeds originate in the Third World. According to these figures, hunger would be eradicated from the world if all rich countries would lower their meat consumption by just 15%! To produce 1 kg beef, 7 kg of grain are needed; 1 kg of chicken requires 2 kg of feed. Rest assured, we are not necessarily advocating a strict vegetarian diet. The agricultural organism needs animals for a sensible nutrient cycle and for fertilization that increases soil fertility. What is crucial is the extent of meat production and the subsequent use of feeds.

In summary, the following results can be compiled: If people in the wealthy countries would consciously give up one "meat day" per week, there would be enough grain to supply all of mankind. It is absolutely not necessary to increase the yield of soils at the expense of biological diversity and soil fertility. As a first line, agricultural goods grown in climatically disadvantaged developing countries should stay within those countries and not be exported as feed anymore. Exports from developing countries such as coffee, cocoa, and tea should be sold to industrial countries at much higher rates and excluding the price boosting wholesale trade, which is in the hands of the industrialized countries. Another direct development aid would be debt relief. This is because poor countries are not able to pay back the loans they have received, especially with interest added. Rather, development aid funds should be gifts.

### Goals of the Economy Block

By selection of one aspect of global economy as an example, it may be possible to expand into a theme that is fascinating as well as disturbing. Depending on the course of the work discussion, a certain market process, such as raw product futures exchange, can be selected and discussed in more detail. It is most important to create intelligence and understanding. On the other side, the problems of global economy should be approached from their roots. The principle of the threefolding of the social organism can be developed from its starting point and with individual examples. In this context it is important to point to existing businesses that develop the threefolding as a social experiment for *real*. Tours of plants and businesses are relevant for grade 12. It makes sense to connect the class with people and organizations who initiate concrete projects with an alternative orientation. Topics should be carefully selected according to people who could be positive examples for the adolescents.

Concrete insights into new achievements of economic control can also be gained in discussions with invited school parents, individuals who are professionally involved in economic processes. In the past, we had the opportunity to speak with the director of a global reinsurance company and with a foreign exchange trader of a Palestine bank based in Zurich, and receive informative reports about the everyday work of these people. If it is possible to invite in a project manager or employee of a so-called foreign aid project for a discussion with your class, it will be most important to find out how realistic their work will help to *eradicate poverty*. Students should find out how to preserve human dignity.

The goal of the entire block is to facilitate this experience for the student: It is possible to work my way into an unknown discipline, understand the processes, and form my own judgments through the subject that I learn about—judgments that do not primarily condemn the work of others but create perspectives for my own responsible actions as an individual and a member of society.

Karst (limestone) Mountains by Li River, Guilin, China. Photo by DSM

# References

Abrams-Planetarium. Abrams *Planetarium Sky Calendar*. East Lansing, MI: Abrams Planetarium.

Adkins, J.F., E.A. Boyle, L. Keigwin, and E. Cortijo. 1997. Variability of the North Atlantic thermohaline circulation during the last interglacial period. *Nature* 390 (6656):154–156.

Agnew, N. and M. Demas. 1998. Preserving the Laetoli footprints. *Scientific American* 279 (3):44–55.

Allegre, C.J. and S.H. Schneider. 1994. The Evolution of the Earth. *Scientific American* 271 (4):66–75.

Anderson, O. 2007. Charles Lyell, uniformitarianism, and interpretive principles. *Zygon* 42 (2):449–462.

Baker, V.R. 1998. Catastrophism and uniformitarianism: logical roots and current relevance in geology. Lyell: *The Past Is the Key to the Present* (143):171–182.

Bambach, R.K., C.R. Scotese, and A.M. Ziegler. 1980. Before Pangea - Geographies of the Paleozoic World. *American Scientist* 68 (1):26–38.

Bangert, John A., Alan D. Fiala, William T. Harris and the United States Naval Observatory. 1991. *Central solar eclipses of 1992: Annular solar eclipse of 4–5 January 1992, total solar eclipse of 30 June 1992, U.S. Naval Observatory circular*. Washington, DC: U.S. Naval Observatory.

Barazangi, M. and J. Dorman. 1969. World Seismicity Maps Compiled from Essa Coast and Geodetic Survey Epicenter Data 1961–1967. *Bulletin of the Seismological Society of America* 59 (1):369–380.

Barker, A.J. 1998. *Introduction to Metamorphic Textures and Microstructures*. 2nd ed. Cheltenham, United Kingdom: Stanley Thornes.

Battan, Louis J. 1974. *Weather*. Englewood Cliffs, NJ: Prentice-Hall.

———. 1979. *Wetter: 13 Tabellen, Geowissen kompakt*. Stuttgart: Enke.

Bauer, Bruce A. 1992. *The Sextant Handbook: Adjustment, Repair, Use and History*. 2nd ed. Camden, ME: International Marine.

Bengtson, Stefan, Veneta Belivanova, Birger Rasmussen and Martin Whitehouse. 2009. The controversial "Cambrian" fossils of the Vindhyan are real but more than a billion years older. *Proceedings of the National Academy of Sciences* 106 (19):7729–7734.

Berner, Georges-Albert, E. Audétat, Alfred Dübendorfer and Hans Rudolf Schmid. 1962. *Pierre-Frédéric Ingold, 1787–1878 / Adolf Guyer-Zeller, 1839–1899*. Vol. 13, *Schweizer Pioniere der Wirtschaft und Technik*. Zürich: Verein für wirtschaftshistorische Studien.

Beste, Dieter. 1996. *Dossier: Klima und Energie, Spektrum der Wissenschaft*. Heidelberg: Spektrum-der-Wiss.-Verl.-Ges.

———. 1997. *Dossier: Welternährung, Spektrum der Wissenschaft*. Heidelberg: Spektrum d. Wissenschaft.

Blanckenburg, Friedhelm von. 1999. Paleoceanography: Tracing Past Ocean Circulation? *Science* 286 (5446):1862b–1863.

Blum, W. 1999. Ozean und Klima. Wehe, wenn der Heizer streikt. *GEO-Wissen Ozeane* 24:45–53.

Blunier, T., J. Chappellaz, J. Schwander, A. Dallenbach, B. Stauffer, T.F. Stocker, D. Raynaud, J. Jouzel, H. B. Clausen, C.U. Hammer and S.J. Johnsen. 1998. Asynchrony of Antarctic and Greenland climate change during the last glacial period. *Nature* 394 (6695):739–743.

Bockemühl, Cornelis Johannes. 1999. Isotopen-Analysen und Altersbestimmungen an Goethes (Urgestein) -Vom Umgang mit den "vier Elementen" in der Geologie. In *Erdentwicklung aktuell erfahren: Geologie und Anthroposophie im Gespräch*, edited by C.J. Bockemühl and C. Ballivet. Stuttgart: Verlag Freies Geistesleben.

Bockemühl, Cornelis Johannes and Christine Ballivet. 1999. *Erdentwicklung aktuell erfahren: Geologie und Anthroposophie im Gespräch*. Stuttgart: Verlag Freies Geistesleben.

Bohnsack, Almut. 1981. *Spinnen und Weben: Entwicklung von Technik und Arbeit im Textilgewerbe*. Orig.-Ausg. ed. *Kulturgeschichte der Naturwissenschaften und der Technik*. Reinbeck bei Hamburg: Rowohlt Taschenbuch-Verl.

———. 2002. *Spinnen und Weben: Entwicklung von Technik und Arbeit im Textilgewerbe, Bramscher Schriften*. Bramsche: Rasch.

Bond, G., W. Showers, M. Cheseby, R. Lotti, P. Almasi, P. deMenocal, P. Priore, H. Cullen, I. Hajdas and G. Bonani. 1997. A pervasive millennial-scale cycle in North Atlantic Holocene and glacial climates. *Science* 278 (5341):1257–1266.

Bosse, Dankmar. 1993. Wie alt ist unser Erde? *Der Merkurstab* Juli/August:382–396.

———. 1995. *Der Mensch in der Erdgeschichte*. Berlin.

———. 1999. Phänomenologischer Geologieunterricht in der Oberstufe. In *Das lebendige Wesen der Erde. Zum Geographieunterricht der Oberstufe*, edited by C. Göpfert. Stuttgart: Freies Geistesleben.

Brasier, M., N. McLoughlin, O. Green and D. Wacey. 2006. A fresh look at the fossil evidence for early Archaean cellular life. *Philosophical Transactions of the Royal Society B-Biological Sciences* 361 (1470):887–902.

Brinkmann, Roland. 1990. *Brinkmanns Abriß der Geologie. Bd. 1: Allgemeine Geologie*. 14. neu bearb. Aufl. ed. Stuttgart: Enke.

British Petroleum Company. Statistical Review of World Energy. London: British Petroleum Co.

Broecker, W.S., and G.H. Denton. 1990. What Drives Glacial Cycles. *Scientific American* 262 (1):48–56.

Bromage, Timothy G. and Friedemann Schrenk. 1999. *African Biogeography, Climate Change & Human Evolution, The human evolution series*. New York: Oxford University Press.

Budd, Graham E. 2008. The earliest fossil record of the animals and its significance. *Philosophical Transactions of the Royal Society B: Biological Sciences* 363 (1496):1425–1434.

Budlong, John P. *Sky and Sextant: Practical Celestial Navigation* (2d). Van Nostrand Reinhold 1978.

Burckhardt, Johann Jakob. 1988. *Die Symmetrie der Kristalle: von René-Just Haüy zur kristallographischen Schule in Zürich*. Basel: Birkhäuser.

Burnham, Dorothy K. *The Comfortable Arts: Traditional Spinning and Weaving in Canada*. National Gallery of Canada 1981.

Burns, J.O., N. Duric, G.J. Taylor,and S.W. Johnson. 1990. Observatories on the Moon. *Scientific American* 262 (3):42–49.

Butterfield, Nicholas J. 2007. Macroevolution and Macroecology through Deep Time. *Palaeontology* 50 (1):41–55.

Canfield, D.E. 2005. The Early History of Atmospheric Oxygen: Homage to Robert M. Garrels. *Annual Review of Earth and Planetary Sciences* 33 (1):1–36.

Chatterjee, Sanat. 2008. *Crystallography and the World of Symmetry*. New York: Springer.

Closs, H., Peter Giese and V. Jacobshagen. 1987. Alfred Wegeners Kontinentalverschiebung aus heutiger Sicht. In *Ozeane und Kontinente: ihre Herkunft, ihre Geschichte und Struktur*, edited by P. Giese. Heidelberg: Spektrum-der-Wissenschaft-Verl.-Ges.

Clottes, Jean. 1996. Meisterschaft aus dem Nichts. "Du."*Die Zeitschrift der Kultur* Heft 8, Aug. 1996:66–71.

———. 2008. *Cave Art*. London; New York: Phaidon Press.

Cloud, P. 1983. The Biosphere. *Scientific American* 249 (3):176–189.

Dalziel, I.W.D. 1995. Earth before Pangea. *Scientific American* 272 (1):58–63.

Damasio, A.R. and H. Damasio. 1992. Brain and Language. *Scientific American* 267 (3):89–95.

Davidson, Norman. 2004. *Sky Phenomena: A Guide to Naked-Eye Observation of the Stars: With Sections on Poetry in Astronomy, Constellation Mythology, and the Southern Hemisphere Sky.* Rev. ed. Hudson, NY: Lindisfarne Books.

Davis, G.R. 1990. Energy for Planet Earth. *Scientific American* 263 (3):54–62.

De Graef, Marc, and Michael E. McHenry. 2007. *Structure of Materials: An Introduction to Crystallography, Diffraction and Symmetry.* Cambridge: Cambridge University Press.

De Gregorio, Bradley T., Thomas G. Sharp, George J. Flynn, Sue Wirick and Richard L. Hervig. 2009. Biogenic origin for Earth's oldest putative microfossils. *Geology 37* (7):631–634.

de Ridder, N., H. Breman, H. van Keulen and T.J. Stomph. 2004. Revisiting a 'cure against land hunger': soil fertility management and farming systems dynamics in the West African Sahel. *Agricultural Systems* 80 (2):109–131.

Defant, Albert. 1958. *Ebb And Flow; The Tides of Earth, Air, and Water, Ann Arbor science library.* Ann Arbor: University of Michigan Press.

———. 1973. *Ebbe und Flut des Meeres, der Atmosphäre und der Erdfeste.* 2. Aufl. ed. Verständliche Wissenschaft, Berlin: Springer-Verl.

Dietrich, Günter and Johannes Ulrich. 1968. *Atlas zur Ozeanographie, BI-Hochschultaschenbücher.* Mannheim: Bibliogr. Inst.

Dones, Roberto. 2003. ecoinvent 2000— Überarbeitung und Ergänzung der Ökoinventare für Energiesysteme. Ittigen: Bundesamt für Energie BFE, Paul Scherrer Institut.

Dreibus-Kapp, G. and L. Schultz. 1999. Chemismus und Bildung des Erdmondes. *Sterne und Weltraum* 9/99:742–753.

Dreibus, G. and H. Wanke. 1990. Comparison of the Chemistry of Moon and Mars. *Smaller Solar System Bodies and Orbits* 10 (3–4):7–16.

Duke, M.B., L.R. Gaddis, G.J. Taylor and H.H. Schmitt. 2006. Development of the moon. *New Views of the Moon* 60:597–655.

Dullo, Wolf-Christian. 1999. Aktuelle Erfahrungen als Verständnisschlüssel für vergangene Entwicklungen - Aktualismus und Uniformitarismus. In *Erdentwicklung aktuell erfahren: Geologie und Anthroposophie im Gespräch*, edited by C.J. Bockemühl and C. Ballivet. Stuttgart: Verlag Freies Geistesleben.

Endres, Klaus-Peter and Wolfgang Schad. 2002. *Moon Rhythms in Nature: How Lunar Cycles Affect Living Organisms.* Edinburgh: Floris Books.

Energiesparende Gebäudesysteme in der Schweiz. 1980. Energiesparende Gebäudesysteme in der Schweiz. Wettewerbs-dokumentation des 50-jährigen Jubiläums. Dübendorf, Switzerland: Oertli AG.

English, Walter. 1969. *The textile industry: an account of the early inventions of spinning, weaving, and knitting machines, Industrial archaeology*, Harlow: Longmans.

Erklärung von Bern (Zürich). *Dokumentation.* Zürich: Erklärung von Bern.

European Commission. Directorate-General for Energy and Transport. 2005. Annual Energy and Transport Review. Luxembourg: Office for Official Publications of the European Communities.

European Commission. Directorate-General for Energy and Transport and Ethniko Metsovio Polytechneio (Greece). 2003. European Energy and Transport: Trends to 2030. Luxembourg: Office for Official Publications of the European Communities.

Fischer-Weltalmanach. *Der Fischer Weltalmanach: Zahlen, Daten, Fakten zu Politik, Wirtschaft, Kultur, yearly since 1959, Fischer Taschenbuch.* Frankfurt am Main: Fischer Taschenbuch Verl.

Freie Hochschule für Geisteswissenschaft (Dornach). Mathematisch-astronomische Sektion. Sternkalender: Erscheinungen am Sternenhimmel: mit naturwissenschaftlichen und literarischen Beiträgen. Dornach: Philosophisch-anthroposophischer Verlag.

Frenz, L. 1998. Stammbaum auferstanden aus Fragmenten. *GEO-Wissen: Die Evolution des Menschen*, Sept. 1998:20–38.

Fricke, H. 2001. Coelacanths: a human responsibility. *Journal of Fish Biology* 59:332–338.

Fricke, H. and K. Hissmann. 2000. Feeding ecology and evolutionary survival of the living coelacanth Latimeria chalumnae. *Marine Biology* 136 (2):379–386.

Fricke, H., K. Hissmann, J. Schauer, M. Erdmann, M.K. Moosa and R. Plante. 2000. Conservation - Biogeography of the Indonesian coelacanths. *Nature* 403 (6765):38–38.

Fricke, H., O. Reinicke, H. Hofer and W. Nachtigall. 1987. Locomotion of the Coelacanth Latimeria-Chalumnae in Its Natural-Environment. *Nature* 329 (6137):331–333.

Frisch, Klaus. 1993. Zur geologischen Datierung des Mondaustritts. *Der Merkurstab* Juli/August, 396–400.

Frischknecht, Rolf, Schweiz. Bundesamt für Energiewirtschaft, Nationaler Energie-Forschungs-Fonds (Basel) and Eidgenössische Technische Hochschule (Zürich). 1995. *Ökoinventare für Energiesysteme: Grundlagen für den ökologischen Vergleich von Energiesystemen und den Einbezug von Energiesystemen in Ökobilanzen für die Schweiz.* 2. Aufl. ed. Bern: ENET.

Gaidos, E., T. Dubuc, M. Dunford, P. McAndrew, J. Padilla-Gamino, B. Studer, K. Weersing and S. Stanley. 2007. The Precambrian emergence of animal life: a geobiological perspective. *Geobiology* 5 (4):351–373.

Ganopolski, A., S. Rahmstorf, V. Petoukhov and M. Claussen. 1998. Simulation of modern and glacial climates with a coupled global model of intermediate complexity. *Nature* 391 (6665):351–356.

Garrison, Tom. 2009. *Oceanography: An Invitation to Marine Science.* 7th ed. Belmont, CA: Brooks/Cole.

Giese, Peter. 1987. *Ozeane und Kontinente: ihre Herkunft, ihre Geschichte und Struktur.* 5. Aufl. ed. Heidelberg: Spektrum-der-Wissenschaft-Verl.-Ges.

Göpfert, Christoph and Dankmar Bosse. 1999. *Das lebendige Wesen der Erde: zum Geographieunterricht der Oberstufe.* 1. Aufl. ed. *Menschenkunde und Erziehung,* Stuttgart: Verl. Freies Geistesleben.

Gould, Stephen Jay. 1965. Is Uniformitarianism Necessary? *American Journal of Science* 263 (3):223–&.

———. 1987. *Time's arrow, time's cycle: myth and metaphor in the discovery of geological time, The Jerusalem-Harvard lectures.* Cambridge, MA: Harvard University Press.

———. 1989. *Wonderful Life: The Burgess Shale and the Nature of History.* 1st ed. New York: W.W. Norton.

Graedel, T.E. and P.J. Crutzen. 1989. The Changing Atmosphere. *Scientific American* 261 (3):58–68.

Graedel, T.E. and Paul J. Crutzen. 1993. *Atmospheric Change: An Earth System Perspective,* New York: W.H. Freeman.

———. 1995. *Atmosphere, climate, and change, Scientific American Library series no. 55.* New York: Scientific American Library; Distributed by W.H. Freeman.

Grant, Norman. 1996. Radioactivity in the History of the Earth. *Archetype* 2.

———. 1999. Radioaktivität in der Erdgeschichte. *In Erdentwicklung aktuell erfahren: Geologie und Anthroposophie im Gespräch,* edited by C.J. Bockemühl and C. Ballivet. Stuttgart: Verlag Freies Geistesleben.

Griffith-Observatory. Sky Calendar & Astro-handbook. Los Angeles: Griffith Observatory.

Groves, D.I., J.S.R. Dunlop, and R. Buick. 1981. An Early Habitat of Life. *Scientific American* 245 (4):64–73.

Haag, Andreas. 1998. Mensch und Kosmos. Aus dem Astronomieunterricht in der 11. Klasse. *Erziehungskunst* 1/98:43–57.

Haldimann, Urs, Stephan Dietrich and Erklärung von Bern (Zürich). 1992. *Unser täglich Fleisch: So Essen wir die Welt Kaputt.* Zürich: Unionsverlag.

Hansen, J. and S. Lebedeff. 1988. Global Surface Air Temperatures - Update through 1987. *Geophysical Research Letters* 15 (4):323–326.

Harrison, Ann E. and ebrary Inc. 2007. *Globalization and Poverty*. Chicago: University of Chicago Press.

Hayes, J.M. 1996. The earliest memories of life on earth. *Nature* 384 (6604):21–22.

Hecht, Ann. 2001. *The Art of the Loom: Weaving, Spinning and Dyeing across the World*. London: British Museum.

Heemstra, P.C., K. Hissmann, H. Fricke, M.J. Smale and J. Schauer. 2006. Fishes of the deep demersal habitat at Ngazidja (Grand Comoro) Island, Western Indian Ocean. *South African Journal of Science* 102 (9–10):444–460.

Heimlicher, M., M. Rauber, Artur Wellinger, Infosolar (Tänikon), Schweizerische Vereinigung für Sonnenenergie und Büro n+1. 1991. *Solarenergie: Basisinformation, Animationsvorschläge und Arbeitsblätter für die praktische Unterrichtsgestaltung mit den Schwerpunkten Energiesparen, passive und aktive Solarenergie, Photovoltaik, Biogas, Holz: Lehrerdokumentation. 3, durchgesehene Aufl.* ed. Brugg: Infosolar.

Held, Wolfgang. 1999. *Die Sonnenfinsternis am 11. August 1999. 4. Aufl.* ed. Stuttgart: Verl. Freies Geistesleben.

Hentschel, Kurt. 1975. *Wolle Spinnen mit Herz und Hand*. Winterbach-Manolzweiler: Webe im Verl.

Herrmann, Albert Günter. 1983. *Radioaktive Abfälle: Probleme und Verantwortung*. Berlin [West]; Heidelberg; New York: Springer.

Herrmann, Albert Günter and Bernhard Knipping. 1993. *Waste disposal and evaporites: contributions to long-term safety*, Lecture notes in *Earth Sciences* 45. Berlin; New York: Springer-Verlag.

Herterich, K. 1987. Die astronomische Theorie der Eiszeiten. *Sterne und Weltraum* 5:272–276.

Hess, Harry Hammond 1962. History of Ocean Basins. In *Petrologic studies; a volume in honor of A.F. Buddington*, edited by Geological Society of America and A.E.J. Engel. New York: Geological Society of America.

Heydebrand, Caroline von. 1984. *The Curriculum of the First Waldorf School*. Forest Row, East Sussex: Steiner Schools Fellowship, 1966 (1984).

Hillebrand, E. 2008. The global distribution of income in 2050. *World Development* 36 (5):727–740.

Hoekman, Bernard M. and M. Olarreaga. *Global Trade and Poor Nations: The Poverty Impacts and Policy Implications of Liberalization*. Brookings Institution Press 2007.

Houghton, R.A. and G.M. Woodwell. 1989. Global Climatic-Change. *Scientific American* 260 (4):36–44.

Howarth, M.K. 1981. Palaeogeography of the Mesozoic. In *The Evolving Earth. Chance, Change & Challenge*, edited by L.R.M. Cocks. London, New York: British Museum (Natural History), Cambridge University Press.

Hublin, J.J. 2009. The origin of Neandertals. *Proceedings of the National Academy of Sciences of the United States of America* 106 (38):16022–16027.

Hublin, J.J., F. Spoor, M. Braun, F. Zonneveld and S. Condemi. 1996. A Late Neanderthal associated with Upper Palaeolithic artefacts. *Nature* 381 (6579):224–226.

Infoenergie. 1994. *Graue Energie, Infoenergie, Publikation der Forschungsanstalt ETH Zürich*. CH-Tänikon: Forschungsanstalt ETH Zürich, Tänikon.

Infosolar. Informations- und Beratungsstelle Sonnenenergie. 1991. *Sonnenenergie und Technik. Fakten, Stichworte, Erklärungen*. Brugg: Informations- und Beratungsstelle Sonnenenergie.

Isbister, John. *Promises Not Kept: Poverty and the Betrayal of Third World Development* (7th). Kumarian Press 2006.

Javaux, Emmanuelle J. 2007. The Early Eukaryotic Fossil Record.

Jensen, S., J.G. Gehling and M.L. Droser. 1998. Ediacara-type fossils in Cambrian sediments. *Nature* 393 (6685):567–569.

Johanson, Donald C., Lenora Carey Johansen and Blake Edgar. *Ancestors: In Search of Human Origins* (1st). Villard Books 1994.

Johanson, Donald C. and Kate Wong. 2009. *Lucy's Legacy: The Quest for Human Origins*. 1st ed. New York: Harmony Books.

Jones, P.D. and T.M.L. Wigley. 1990. Global Warming Trends. *Scientific American* 263 (2):84–91.

Joseph, B.D. and S.S. Mufwene. 2008. Parsing the evolution of language. *Science* 320 (5875):446–446.

Jouzel, J., V. Masson-Delmotte, O. Cattani, G. Dreyfus, S. Falourd, et al. 2007. Orbital and millennial Antarctic climate variability over the past 800,000 years. *Science* 317 (5839):793–796.

Kahmen, Heribert and Wolfgang Faig. 1988. *Surveying*, Berlin: de Gruyter.

Karutz, Matthias. 1991. Zur Betrachtung technischer Entwicklungen. Technologie-Unterricht in der 10. Klasse. *Erziehungskunst* 1/91:10–26.

———. 1992. *Zur Lebenskunde III. Die Spinnepoche.* Stuttgart: Pädagogische Forschungsstelle beim Bund der Freien Waldorfschulen, Stuttgart.

Keeling, Ralph F. *The Atmosphere* (1st). Elsevier 2006.

Keidel, Claus G. and Raymund Windolf. 1986. *Wolkenbilder - Wettervorhersage: mit 104 Farbfotos von typischen Wettererscheinungen.* 3. durchges. Aufl. ed. BLV-Naturführer, München: BLV-Verlagsgesellschaft.

Kellogg, L.H., B.H. Hager and R.D. van der Hilst. 1999. Compositional stratification in the deep mantle. *Science* 283 (5409):1881–1884.

Kelts, Kerry R. 1978. Geological and Sedimentary Evolution of Lakes Zurich and Zug, Switzerland. Diss ETH Zürich, 1978.

Kleber, Will, Hans-Joachim Bautsch, Irmgard Kleber and Joachim Bohm. 1990. *Einführung in die Kristallographie.* 17, stark bearb. Aufl. / von Hans-Joachim Bautsch, ed. Berlin: Verl. Technik.

Knoll, A.H., E.J. Javaux, D. Hewitt and P. Cohen. 2006. Eukaryotic organisms in Proterozoic oceans. *Philosophical Transactions of the Royal Society B: Biological Sciences* 361 (1470):1023–1038.

Knoll, Andrew, Malcolm Walter, Guy Narbonne and Nicholas Christie-Blick. 2006. The Ediacaran Period: a new addition to the geologic time scale. *Lethaia* 39:13–30.

Knutsen, H.M. 2010. Hanging by a Thread: Cotton, Globalization and Poverty in Africa. *Professional Geographer* 62 (1):147–149.

Koehler, Reinhard. 1986. Stauen und Strömen im Weltmeer. In *Der Organismus der Erde: Grundlagen einer neuen Ökologie*, edited by B. Endlich and J. Bockemühl. Stuttgart: Verl. Freies Geistesleben.

Kohl, Richard and Organisation for Economic Cooperation and Development. *Globalisation, poverty and inequality.* Development Centre of the Organisation for Economic Cooperation and Development 2003.

Kracht, Uwe. 1999. *Food Security and Nutrition: The Global Challenge* /ed. by Uwe Kracht, *Spektrum.* Münster [u.a.]: Lit [u.a.].

Kranich, Ernst Michael. 1997. *Pflanze und Kosmos: Grundlinien einer kosmologischen Botanik.* Neuausg. (3. aktualisierte und erw. Aufl.) ed. Stuttgart: Verl. Freies Geistesleben.

Kremer, B.P. 1998. Erste moderne Tiere schon im Präkambrium. *Spektrum der Wissenschaft* Juni 1998:27–31.

Kübler, Fritz. 2000. Elemente der Wirtschaft. In *Geographie - Wirtschaft -Technik und das soziale Leben der Gegenwart. Anregungen für die 11. Klasse. Aus dem fächerübergreifenden Oberstufenunterricht der Waldorfschulen,* edited by F. Kübler. Kassel: Pädagogische Forschungsstelle beim Bund der Freien Waldorfschulen, Abt. Kassel.

———. 2000. *Geographie - Wirtschaft - Technik und das soziale Leben der Gegenwart. Anregungen für die 11. Klasse, Aus dem fächerübergreifenden Oberstufenunterricht der Waldorfschulen.* Kassel: Bildungswerk, Lehrmittelabteilung, Brabanterstrasse 45, D-34131 Kassel.

Kuhn, Thomas S. 1957. *The Copernican Revolution: Planetary Astronomy in the Development of Western Thought.* Cambridge: Harvard University Press.

Kukathas, Uma. 2009. *The Global Food Crisis, Current Controversies.* Detroit, MI: Greenhaven Press.

Lal, R. 2008. Sustainable horticulture and resource management. *Proceedings of the International Symposium on Sustainability through Integrated and Organic Horticulture* (767):19–44.

Lalli, Carol M. and Timothy Richard Parsons. 1997. *Biological Oceanography: An Introduction*. 2nd ed., Open University oceanography series. Oxford, England: Butterworth Heinemann.

Lang, Kenneth R. and Charles Allen Whitney. 1991. *Wanderers In Space: Exploration and Discovery in the Solar System*. Cambridge; New York: Cambridge University Press.

Laskar, J., F. Joutel,and P. Robutel. 1993. Stabilization of the Earth's Obliquity by the Moon. *Nature* 361 (6413):615–617.

Leakey, M. and A. Walker. 1997. Early hominid fossils from Africa. *Scientific American* 276 (6):74–79.

Leakey, Richard E. 1994. *The Origin of Humankind, Science Masters Series*. New York: BasicBooks.

Leakey, Richard E. and Roger Lewin. 1992. *Origins Reconsidered: In Search of What Makes Us Human*. New York: Doubleday.

———. *The Sixth Extinction: Patterns of Life and the Future of Humankind* (1st). Doubleday 1995.

Lehman, S. 1997. Climate change - Sudden end of an interglacial. *Nature* 390 (6656):117–&.

Lehrs, Ernst. 1953. *Mensch und Materie: e. Beitr. zur Erweiterung d. Naturerkenntnis nach d. Methode Goethes*. Frankfurt am Main: Klostermann.

———. *Man or Matter: Introduction to a Spiritual Understanding of Nature on the Basis of Goethe's Method of Training Observation and Thought* (2d). Faber and Faber 1958.

Levinton, J., L. Dubb and G.A. Wray. 2004. Simulations of evolutionary radiations and their application to understanding the probability of a Cambrian explosion. *Journal of Paleontology* 78 (1):31–38.

Levinton, J.S. 1992. The Big-Bang of Animal Evolution. *Scientific American* 267 (5):84–91.

———. 2008. The Cambrian Explosion: How Do We Use the Evidence? *Bioscience* 58 (9):855–864.

Lewin, Roger. 1993. T*he origin of modern humans, Scientific American Library series*. New York: Scientific American Library: Distributed by W.H. Freeman.

———. 2005. *Human Evolution: An Illustrated Introduction*. 5th ed. Malden, MA: Blackwell Pub. Co.

Lewin, Roger and Robert Foley. 2004. *Principles of Human Evolution*. 2nd ed. Malden, MA: Blackwell Pub. Co.

Lines, Thomas. 2008. *Making Poverty: A History*. London; New York: Zed Books, Distributed in the USA by Palgrave Macmillan.

Little, Daniel. *The Paradox of Wealth and Poverty: Mapping the Ethical Dilemmas of Global Development*. Westview Press 2003.

Liu, J.G., A.J.B. Zehnder and H. Yang. 2009. Global consumptive water use for crop production: The importance of greenwater and virtual water. *Water Resources Research* 45: W05428, 15 pp.

Loske, Reinhard,and Bund für Umwelt und Naturschutz Deutschland. 1996. *Zukunftsfähiges Deutschland: ein Beitrag zu einer global nachhaltigen Entwicklung; Studie des Wuppertal Instituts für Klima, Umwelt, Energie GmbH*. 3. [Dr.]. ed. Basel.

Lovejoy, C.O. 1988. Evolution of Human Walking. *Scientific American* 259 (5):118–125.

Lovejoy, C. Owen, Gen Suwa, Linda Spurlock, Berhane Asfaw and Tim D. White. 2009. The Pelvis and Femur of Ardipithecus ramidus: The Emergence of Upright Walking. *Science* 326 (5949):71–716.

Marshall, Charles R. 2006. Explaining the Cambrian "Explosion" of Animals. *Annual Review of Earth and Planetary Sciences* 34 (1):355–384.

Martin, W. 2005. Archaebacteria (Archaea) and the origin of the eukaryotic nucleus. *Current Opinion in Microbiology* 8 (6):630–637.

Martin, W. and M. Müller. 1998. The hydrogen hypothesis for the first eukaryote. *Nature* 392 (6671):37–41.

May, Markus and Jenni Energietechnik (Oberburg). 1996. *Graue Energie und Umweltbelastungen von Heizungssystemen*. Oberburg: Jenni Energietechnik.

McMenamin, M.A.S. 1987. The Emergence of Animals. *Scientific American* 256 (4):94–102.

McMenamin, M.A.S. 2000. The crucible of creation: The Burgess Shale and the rise of animals. *Notes and Records of the Royal Society of London* 54 (3):407–408.

Mechsner, F. 1998. Wer sprach das erste Wort? *GEO-Wissen* Sept. 1998:78–83.

Moorbath, S. 1977. Oldest Rocks and Growth of Continents. *Scientific American* 236 (3):92–104.

———. 2009. The discovery of the earth's oldest rocks. *Notes and Records of the Royal Society* 63 (4):381–392.

Müller, Helmut. 1971. *Astronomische Orts-, Zeit- und Azimutbestimmungen mit dem Kern DKM 3-A.* Aarau: Kern.

Murphy, J.B., and R.D. Nance. 1992. Mountain Belts and the Supercontinent Cycle. *Scientific American 266* (4):84–91.

Nance, R.D., T.R. Worsley and J.B. Moody. 1988. The Supercontinent Cycle. *Scientific American* 259 (1):72–79.

Narbonne, Guy M. 2005. The Ediacara Biota: Neoproterozoic Origin of Animals and Their Ecosystems. *Annual Review of Earth and Planetary Sciences* 33 (1):421–442.

Ohlendorf, H.-Christian. 1994. *Feldmessen.* Kassel: Pädagogische Forschungsstelle, Bund der Freien Waldorfschulen.

Owen, H.G. 1981. Constant dimensions of an expanding earth. In *The Evolving Earth. Chance, Change & Challenge*, edited by L.R.M. Cocks. London, New York: British Museum (Natural History); Cambridge University Press.

———. 1983. *Atlas of Continental Displacement: 200 Million Years to the Present.* Cambridge: Cambridge University Press.

Paillard, D. 1998. The timing of Pleistocene glaciations from a simple multiple-state climate model. *Nature* 391 (6665):378–381.

Parker, Robert Lüling and Hans Ulrich Bambauer. 1975. *Mineralienkunde: ein Leitfaden für den Sammler.* 5. Aufl. ed. Thun: Ott.

Pavoni, N. 1991. Bipolarity in Structure and Dynamics of the Earth's Mantle. *Eclogae Geologicae Helvetiae* 84 (2):327–343.

———. 1997. Geotectonic bipolarity - evidence of bicellular convection in the earth's mantle. *South African Journal of Geology* 100 (4):291–299.

Pavoni, N. and M.V. Muller. 2000. Geotectonic bipolarity, evidence from the pattern of active oceanic ridges bordering the Pacific and African plates. *Journal of Geodynamics* 30 (5):593–601.

Perez-Malvaez, C., B.H. Alfredo, F.O. Manuel and R.R. Rosaura. 2006. Ninety-four years of the theory of the continental drift of Alfred Lothar Wegener. *Interciencia* 31 (7):536–543.

Pfiffner, Othmar Adrian and Nationales Forschungsprogramm 20 Geologische Tiefenstruktur der Schweiz. 1997. *Deep structure of the Swiss Alps: Results of NRP 20.* Basel [etc.]: Birkhäuser Verlag.

Pflug, Hans D. 1984. Early Geological Record and the Origin of Life. *Naturwissenschaften* 71 (2):63–68.

———. 1989. *Fossilien: Bilder Frühen Lebens.* Heidelberg: Spektrum d. Wiss.

———. 2001. Earliest organic evolution. Essay to the memory of Bartholomew Nagy. *Precambrian Research* 106 (1–2):79–91.

Pichler, Hans. 1988. *Vulkanismus: Naturgewalt, Klimafaktor und kosmische Formkraft.* 2. Aufl. ed. Heidelberg: Spektrum d. Wiss.

Pimentel, David and Marcia Pimentel. 2008. *Food, Energy, and Society.* 3rd ed. Boca Raton, FL: CRC Press.

Press, Frank and Raymond Siever. 2001. *Understanding Earth.* 3rd ed. New York: W.H. Freeman.

———. 2003. *Allgemeine Geologie: Einführung in das System Erde.* 3. Aufl. ed. Heidelberg: Spektrum, Akad. Verl.

Rasmussen, Birger, Ian R. Fletcher, Jochen J. Brocks and Matt R. Kilburn. 2008. Reassessing the first appearance of eukaryotes and cyanobacteria. *Nature* 455 (7216):1101–1104.

Reif, W.E. 2000. Darwinism, gradualism and uniformitarianism. *Neues Jahrbuch für Geologie und Palaontologie-Monatshefte* (11):669–680.

Riebesell, U. 2008. Climate change - Acid test for marine biodiversity. *Nature* 454 (7200):46–47.

Riebesell, U. and D. Wolfgladrow. 1992. Das Defizit in der Kohlenstoffbilanz. *Spektrum der Wissenschaft* 7:28–32.

Riebesell, U., D.A. Wolfgladrow and V. Smetacek. 1993. Carbon-Dioxide Limitation of Marine-Phytoplankton Growth-Rates. *Nature* 361 (6409):249–251.

Rifkin, Jeremy. 1992. *Beyond Beef: The Rise and Fall of the Cattle Culture*. New York: Plume.

Rohrbach, Klaus. 1999. Die Erde als Ganzes - ein lebendiger Organismus. In *Das lebendige Wesen der Erde: zum Geographieunterricht der Oberstufe*, edited by C. Göpfert and D. Bosse. Stuttgart: Verl. Freies Geistesleben.

———. 2000. Keine Lehrplanangaben Steiners zur Geographie? In *Geographie -Wirtschaft - Technik und das soziale Leben der Gegenwart. Anregungen für die 11. Klasse. Aus Dem Fächerübergreifenden Oberstufenunterricht der Waldorfschulen*, edited by F. Kübler. Kassel: Pädagogische Forschungsstelle beim Bund der Freien Waldorfschulen, Abt. Kassel.

Root, T.L. and S.H. Schneider. 2006. Conservation and climate change: The challenges ahead. *Conservation Biology* 20 (3):706–708.

Rosslenbroich, B. 2006. The notion of progress in evolutionary biology - the unresolved problem and an empirical suggestion. *Biology & Philosophy* 21 (1):41–70.

———. 2009. The theory of increasing autonomy in evolution: a proposal for understanding macroevolutionary innovations. *Biology & Philosophy* 24 (5):623–644.

Ruddiman, W.F. and J.E. Kutzbach. 1991. Plateau Uplift and Climatic Change. *Scientific American* 264 (3):66–75.

Rykart, Rudolf. 1995. *Quarz-Monographie: die Eigenheiten von Bergkristall, Rauchquarz, Amethyst, Chalcedon, Achat, Opal und anderen Varietäten*. 2. Aufl. ed. Thun: Ott.

Sachs, Wolfgang, Bund für Umwelt und Naturschutz Deutschland and Wuppertal-Institut für Klima Umwelt Energie. 2009. *Zukunftsfähiges Deutschland in einer globalisierten Welt: ein Anstoß zur gesellschaftlichen Debatte; eine Studie des Wuppertal Instituts für Klima, Umwelt, Energie*. 3. Aufl. ed. Fischer; Frankfurt am Main: Fischer-Taschenbuch-Verl.

Sax, Anna. 1997. *[Das Existenzmaximum] [Bildmaterial]: [Grundlagen für eine zukunftsfähige Schweiz]*. Basel: Ökomedia.

Sax, Anna, Peter Haber, Daniel Wiener, Erklärung von Bern (Zürich) and Ökomedia (Basel). 1997. *Das Existenzmaximum: Grundlage für eine zukunftsfähige Schweiz*. Zürich: Werd Verlag.

Schad, Wolfgang. 1985. Gestaltmotive der fossilen Menschenformen. In *Goetheanistische Naturwissenschaft, Bd. 4: Anthropologie*, edited by W. Schad. Stuttgart: Verl. Freies Geistesleben.

———. 1996. Urgeschichte Palästinas. In *Mitte der Erde: Israel und Palästina im Brennpunkt natur- und kulturgeschichtlicher Entwicklungen*, edited by A. Suchantke. Stuttgart: Verlag Freies Geistesleben.

Schmidt, Thomas. 1987. *Zum Astronomie-Unterricht an der Waldorfschule*. Stuttgart: Pädagogische Forschungsstelle beim Bund der Freien Waldorfschulen, Stuttgart.

———. 1998. Die Wissenschaftlichkeit der Astronomie im Unterricht. *Erziehungskunst* 1/98:15–26.

Schmincke, Hans-Ulrich. 2004. *Volcanism*. Berlin: Springer.

———. 2006. *Vulkanismus* Sonderausg. [Nachdr. der] 2., überarb. und erw. Aufl. ed. Darmstadt: Wiss. Buchges.

Schmutz, H. 1987. Sonnenenergie - eine Alternative zur Atomenergie! In *Jahresarbeit 12. Klasse*. Wetzikon, Switzerland: Rudolf Steiner Schule, Zürich-Wetzikon.

Schmutz, Hans-Ulrich. 1976. Der Mafitit-Ultramafitit-Komplex zwischen Chiavenna und Val Bondasca (Provinz Sondrio, Italien; Kt. Graubünden, Schweiz). Zugl: Diss ETH Zürich, 1974, [s.n.], Bern.

———. 1986. *Die Tetraederstruktur der Erde: eine geologisch-geometrische Untersuchung anhand der Plattentektonik*. Stuttgart: Verlag Freies Geistesleben.

———. 1987. *Die Großtektonik der Erde - ihre Tetraedergestalt, vier großformatige Demonstrationsblätter für den Geologie-Unterricht mit Beitext*. Stuttgart: Pädagogische Forschungsstelle des Bund der Freien Waldorfschulen, Abt. Stuttgart.

————. 1993. Erdwesenskunde als Menschenkunde. *Erziehungskunst* Febr. 93:180–196.

————. 1996. Zur Geologie Palästinas. In *Mitte der Erde: Israel und Palästina im Brennpunkt natur- und kulturgeschichtlicher Entwicklungen*, edited by A. Suchantke. Stuttgart: Verlag Freies Geistesleben.

————. 1998. Treibhauseffekt - versiegt der Golfstrom? *info* 3 2/1998:13–17.

————. 1999. Erdwesenskunde als Menschenkunde. In *Das lebendige Wesen der Erde. Zum Geographieunterricht der Oberstufe*, edited by C. Göpfert. Stuttgart: Freies Geistesleben.

————. 2000. Energiewirtschaft I; Astronomie. In *Geographie - Wirtschaft - Technik und das soziale Leben der Gegenwart. Anregungen für die 11. Klasse. Aus Dem fächerübergreifenden Oberstufenunterricht der Waldorfschulen*, edited by F. Kübler. Kassel: Pädagogische Forschungsstelle beim Bund der Freien Waldorfschulen, Abt. Kassel.

————. 2000. Goethes anschauende Urteilskraft, die Systematik der Kristalle und die Signatur der Plattentektonik in der Geologie. In *Goethes Beitrag zur Erneuerung der Naturwissenschaften: das Buch zur gleichnamigen Ringvorlesung an der Universität Bern zum 250, Geburtsjahr Goethes*, edited by P. Heusser and Universität Bern. Kollegiale Instanz für Komplementärmedizin. Bern: Haupt.

————. 2005. *Acht geologische Exkursionen durch die Alpen: vom Beobachten der Gesteine zum Verstehen der Erde*. Stuttgart: Freies Geistesleben.

Schmutz, Hans-Ulrich, Manfred von Mackensen, Klaus Rohrbach and Astrid Lütje. 2004. *Erdgeschichte, Paläontologie und Aspekte der Paläoanthropologie. Materialien zum Erdkundeunterricht in der 12. Klasse, Aus Dem fächerübergreifenden Oberstufenunterricht der Waldorfschulen*. Kassel: Pädagogische Forschungsstelle beim Bund der Freien Waldorfschulen, Abt. Kassel.

Schneider, Götz. 2004. *Erdbeben: [eine Einführung für Geowissenschaftler und Bauingenieure]*. 1. Aufl. ed. München: Elsevier, Spektrum, Akad. Verl.

Schneider, Hermann. 1982. *Der Urknall und die absoluten Datierungen*. [5. Tsd.]. ed, Wort und Wissen, Neuhausen-Stuttgart: Hänssler.

Schneider, S.H. 1987. Climate Modeling. *Scientific American* 256 (5):72–78.

————. 1989. The Changing Climate. *Scientific American* 261 (3):70–79.

————. 2001. What is 'dangerous' climate change? *Nature* 411 (6833):17–19.

Schopf, J.W. 2009. The hunt for Precambrian fossils: An abbreviated genealogy of the science. *Precambrian Research* 173 (1–4):4–9.

Schrenk, Friedemann. 2008. *Die Frühzeit des Menschen: der Weg zum Homo sapiens*. 5, neubearb. und erg. Aufl. ed. München: Verlag C.H. Beck.

Schrenk, Friedemann, Stephanie Müller and Christine Hemm. 2009. *The Neanderthals*. London: Routledge.

Schultz, Joachim. 1977. *Rhythmen der Sterne: Erscheinungen und Bewegungen von Sonne, Mond und Planeten*. [2. durchges. Aufl. fotomech. Nachdr] ed. Dornach: Philosophisch-Anthroposophischer Verlag am Goetheanum.

————. 1986. *Movement and Rhythms of the Stars: A Guide to Naked-Eye Observation of Sun, Moon, and Planets*. Edinburgh; Spring Valley, NY: Floris Books; Anthroposophic Press.

Schwander, J., J.M. Barnola, C. Andrie, M. Leuenberger, A. Ludin, D. Raynaud and B. Stauffer. 1993. The Age of the Air in the Firn and the Ice at Summit, Greenland. *Journal of Geophysical Research-Atmospheres* 98 (D2):2831–2838.

Schweiz. Bundesamt für Energie, and Schweiz Bundesamt für Energiewirtschaft. Schweizerische Gesamtenergiestatistik. Bern: Bundesamt für Energie.

Schweiz. Arbeitsgemeinschaft für forstlichen Straßenbau (SAFS). 1977. *Merkblätter über den Bau und Unterhalt von Wald- und Güterstraßen*. Vol. 102–200. Zürich.

Schweiz. Bundesamt für Konjunkturfragen. Impulsprogramm PACER. 1995. *Solare Warmwassererzeugung: Realisierung, Inbetriebnahme und Wartung*. Bern: Bundesamt für Konjunkturfragen.

Schweiz. Energiestiftung SES. SOFAS. Sonnenenergie Schulungskurs. In *Kurs 81,* edited by SOFAS: Schweiz. Energiestiftung SES.

———. 1983. *Wärme-Kraft-Koppelung in dezentralen Anlagen.* Zürich: Schweiz. Energiestiftung SES.

Schweizerische Vereinigung für Sonnenenergie. 1990. *Solar 91 zum 700. Geburtstag der Schweizerischen Eidgenossenschaft: Solar 91 - Handbuch / Solar 91 für eine Energie-Unabhängigere Schweiz: Für jede Schweizer Gemeinde bis zum Jahr 2000 eine Solaranlage von 1 KW bis 1 MW.*

Shen, Yanan, Tonggang Zhang, and Paul F. Hoffman. 2008. On the coevolution of Ediacaran oceans and animals. *Proceedings of the National Academy of Sciences* 105 (21):7376–7381.

Siever, R. 1983. The Dynamic Earth. *Scientific American* 249 (3):46–55.

Smith, A.G., A.M. Hurley, and J.C. Briden. 1981. Phanerozoic Paleocontinental World Maps. In *Cambridge earth science series.* Cambridge; New York: Cambridge University Press,.

Stalder, Hans Anton. 1998. *Mineralienlexikon der Schweiz eine topographische Mineralogie, alphabetisch geordnet nach den einzelnen Mineralarten, mit zwei Spezialkapiteln "Mineralvorkommmen" und "Schweizer Meteorite," mit.* Basel: Wepf.

Stanley, Steven M. 1985. Climatic Cooling and Plio-Pleistocene Mass Extinction of Mollusks around the Margins of the Atlantic. *South African Journal of Science* 81 (5):266–266.

———. 1987. *Extinction, Scientific American Library.* New York: Scientific American Library, distributed by W.H. Freeman.

———. 2008. *Earth System History.* 3rd ed. New York: W.H. Freeman and Co.

Stanley, Steven M. and L.D. Campbell. 1981. Neogene Mass Extinction of Western Atlantic Mollusks. *Nature* 293 (5832):457–459.

Stanley, Steven M., and M.G. Powell. 2003. Depressed rates of origination and extinction during the late Paleozoic ice age: A new state for the global marine ecosystem. *Geology* 31 (10):877–880.

Stanley, Steven M. and Volker Schweizer. 1994. *Historische Geologie: eine Einführung in die Geschichte der Erde und des Lebens.* Heidelberg: Spektrum Akad. Verl.

Stanley, Steven M., and X. Yang. 1994. A Double Mass Extinction at the End of the Paleozoic Era. *Science* 266 (5189):1340–1344.

Stauffer, B. 1993. The Greenland Ice Core Project. *Science* 260 (5115):1766–1767.

Stauffer, B., and T. Stocker. 1995. Climate instabilities from Greenland records. *Ocean and the Poles* 381:253–261.

Steffensen, J.P., K.K. Andersen, M. Bigler, H.B. Clausen, D. Dahl-Jensen, H. Fischer, K. Goto-Azuma, M. Hansson, S.J. Johnsen, J. Jouzel, V. Masson-Delmotte, T. Popp, S.O. Rasmussen, R. Rothlisberger, U. Ruth, B. Stauffer, M.L. Siggaard-Andersen, A.E. Sveinbjornsdottir, A. Svensson and J.W.C. White. 2008. High-resolution Greenland Ice Core data show abrupt climate change happens in few years. *Science* 321 (5889):680–684.

Steiner, Rudolf. 1884–1897. *Einleitungen zu Goethes Naturwissenschaftlichen Schriften. Zugleich eine Grundlegung der Geisteswissenschaft (Anthroposophie).* GA 1, translation: 1988. Goethean science: Mercury Press.

———. 1886. *Grundlinien einer Erkenntnistheorie der Goetheschen Weltanschauung, mit besonderer Rücksicht auf Schiller.* GA 2, translation: Steiner, Rudolf and William Lindeman. 1988. *The Science of Knowing: Outline of an Epistemology Implicit in the Goethean WorldView: with Particular Reference to Schiller.* Spring Valley: Mercury Press.

———. 1894. *Philosophie der Freiheit. Grundzüge einer modernen Weltanschauung - Seelische Beobachtungsresultate nach naturwissenschaftlicher Methode.* GA 4. translation: 1999. *The Philosophy of Freedom (The Philosophy of Spiritual Activity): The Basis for a Modern World Conception: Some Results of Introspective Observation Following the Methods of Natural Science.* [Rev. new ed. London: R. Steiner Press.

———. 1904. *Theosophie. Einführung in übersinnliche Welterkenntnis und*

*Menschenbestimmung*. Tb 615 from GA 9. translation: 1994. *Theosophy: An Introduction to the Spiritual Processes in Human Life and in the Cosmos*, Classics in anthroposophy. Hudson, NY: Anthroposophic Press.

———. 1905a. *Geisteswissenschaft und soziale Frage. Die wahren Gesetze des menschlichen Zusammenlebens*. Excerpts from GA 34, translation: 1964. *The Nature of Anthroposophy*. 1st ed, Free Deeds Books. Blauvelt, NY: Rudolf Steiner Publications. also translated as: *Anthroposophy and the Social Question*, first published in 1982 by Mercury Press, and *Spiritual Science and the Social Question*, published 1958 by Anthroposophic Press, both available online at http://rsarchive.org.

———. 1905b. *Grundelemente der Esoterik*. GA 93a, translation: 1983. *Foundations of Esotericism: notes of an esoteric course in the form of thirty-one lectures held in Berlin from 26th September to 5th November 1905*. London: Rudolf Steiner Press.

———. 1910a. *Die Geheimwissenschaft im Umriss*. GA 13, translation: 1979. *Occult Science: An Outline*. London: Rudolf Steiner Press.

Steiner, Rudolf and Catherine E. Creeger. 1997. *An Outline of Esoteric Science*, Classics in Anthroposophy. Hudson, NY: Anthroposophic Press.

———. 1910b. *Die Mission einzelner Volksseelen im Zusammenhang mit der germanisch-nordischen Mythologie*, lecture cycle 6/7–17/1910, GA 121. translation: 1989. *The Mission of Folk-souls (in connection with Germanic-Scandinavian mythology): a course of eleven lectures given at Oslo, Norway, in June 1910*. 1st American rev. ed. Blauvelt, NY: Spiritual Research Editions.

———. 1917. *Von Seelenrätseln*. Tb 637 from GA 21. partially translated: Steiner, Rudolf and Owen Barfield. 1970. *The Case for Anthroposophy: being extracts from von Seelenrätseln*. London: Rudolf Steiner Press.

———. 1919a. *Allgemeine Menschenkunde als Grundlage der Pädagogik (I)*. GA 293. translation: Steiner, Rudolf and Robert F. Lathe. 1996. *The Foundation of Human Experience*. Hudson, NY: Anthroposophic Press.

———. 1919b. *Erziehungskunst. Methodisch-Didaktisches (II)*. GA 294. translation: 2000. *Practical Advice to Teachers, Foundations of Waldorf Education 2*. Great Barrington, MA: Anthroposophic Press.

———. 1919c. *Erziehungskunst Seminarbesprechungen und Lehrplanvorträge (III)*. GA 295. translation: 1997. *Discussions with Teachers: fifteen discussions with the teachers of the Stuttgart Waldorf School, August 21–September 6, 1919: three lectures on the curriculum, September 6, 1919, Foundations of Waldorf education 3*. Hudson, NY: Anthroposophic Press.

———. 1919d. *"Drei Vorträge über Volkspädagogik,"* 5/11, 5/18, and 6/1 1919, in: *Geisteswissenschaftliche Behandlung sozialer und pädagogischer Fragen*. GA 192. translation: 1997. *Education as a Force for Social Change, Foundations of Waldorf education 4*. Hudson, NY: Anthroposophic Press.

———. 1919e. *Die Kernpunkte der sozialen Frage in den Lebensnotwendigkeiten der Gegenwart und Zukunft*. GA 23. translation: 1985. *The Renewal of the Social Organism*. Spring Valley, NY; London: Anthroposophic Press; Rudolf Steiner Press.

———. 1921. *Das Verhältnis der verschiedenen naturwissenschaftlichen Gebiete zur Astronomie. Dritter naturwissenschaftlicher Kurs: Himmelskunde in Beziehung zum Menschen und zur Menschenkunde*. 1/1–1/18/1921, GA 323. translation: Steiner, Rudolf and David Eyes. 2010. *The Relation of the Diverse Branches of Natural Science to Astronomy* (eighteen lectures January 1–18, 1921). Online at: http://awakenings.com/jcms/anthroposophy-and-goethean/45-rudolf-steiner-third-cycle.html.

———. 1922a. *Erziehungsfragen im Reifealter. Zur künstlerischen Gestaltung des Unterrichts*, zwei Vorträge: 6/21 a. 6/22/1922. Excerpt from GA 302a, translation: *Education for Adolescents*. Only known as a free online source at the Rudolf Steiner Archive: http://wn.rsarchive.org/Education/EduAdo_index.html.

———. 1922b. *Nationalökonomischer Kurs. Nationalökonomisches Seminar.* GA Tb 731 (1 volume from GA 340 and 341). translation: 1996. *Economics: The World as One Economy.* Reprinted ed. Bristol: New Economy Publications.

———. 1924a. *Gestalt und Entstehung der Erde und des Mondes.* Vulkanismus, Vortrag 18.9.24, in: Die Schöpfung der Welt und des Menschen. Erdenleben und Sternenwirken. GA 354. translation: 1987. *The Evolution of Earth* and *Man and Influence of the Stars*: 14 lectures to the workmen, June–Sept., 1924 in Dornach. Hudson, NY: Anthroposophic Press.

———. 1924b. *Konferenzen mit den Lehrern der Freien Waldorfschule 1919 bis 1924*, 3 Bde, GA 300 a–c, translation: 1998. *Faculty Meetings with Rudolf Steiner, Foundations of Waldorf Education 8.* Hudson, NY: Anthroposophic Press.

Stocker, T.F. and A. Schmittner. 1997. Influence of $CO_2$ emission rates on the stability of the thermohaline circulation. *Nature* 388 (6645):862–865.

Stockmeyer, E.A. Karl. 1985. *Rudolf Steiner's Curriculum for Waldorf Schools*: Forest Row, East Sussex: Steiner Schools Fellowship, 1969 (1985).

Strahm, Rudolf H. 1992. *Wirtschaftsbuch Schweiz: Das moderne Grundwissen über Oekonomie und Oekologie: Ein Arbeits-, Lehr-, Lern- und Informationsbuch mit Schaubildern und Kommentaren.* 3. neu bearbeitete und aktualisierte Aufl. ed. Aarau: Sauerländer.

———. 1995. *Warum sie so arm sind: Arbeitsbuch zur Entwicklung der Unterentwicklung in der Dritten Welt mit Schaubildern und Kommentaren.* [9. Aufl.] ed. Wuppertal: Hammer.

———. 2008. *Warum wir so reich sind: Wirtschaftsbuch Schweiz.* Bern: h.e.p.-Verlag.

———. 2008. *Warum wir so reich sind [Elektronische Daten].* Bern: h.e.p.

Streit, Bruno. 1995. *Evolution des Menschen.* Heidelberg: Spektrum, Akad. Verl.

Strobach, Klaus. 1990. Vom *"Urknall" zur Erde: Werden und Wandlung unseres Planeten im Kosmos.* Augsburg: Weltbild-Verl.

Suchantke, Andreas. 1967. Paläontologie als Menschenkunde. *Die Menschenschule: allgemeine Erziehungskunst im Sinne Rudolf Steiners*, April/Mai 1967:105–147.

———. 1996. *Mitte der Erde: Israel und Palästina im Brennpunkt natur- und kulturgeschichtlicher Entwicklungen.* [2. überarb. Aufl.] ed. Stuttgart: Verlag Freies Geistesleben.

———. 2001. *Eco-Geography, Renewal in Science.* Great Barrington, MA: Lindisfarne Books.

TA *Tages-Anzeiger*, Swiss daily newspaper, ed. By TA Media AG, Zürich. Articles from 11/27/1997, 77–88: special on climate; 5/3/1996, 82: Who sows energy should also harvest it; 8/19/1994, 41: We consume energy on tick; 6/16/1994: Soon much cheaper and more efficient solar panels; 3/14/1994, 84: Live comfortably even without heating; 7/25/1986, 2: Getting off until 2020.

Tansey, Geoff and Joyce D'Silva. 1999. *The Meat Business: Devouring a Hungry Planet.* New York: St. Martin's Press.

Tattersall, Ian. 2000. Once we were not alone. *Scientific American* 282 (1):56–62.

———. 2009. *The Fossil Trail: How We Know What We Think We Know about Human Evolution.* 2nd ed. New York: Oxford University Press.

Tattersall, Ian and J.H. Schwartz. 2009. Evolution of the Genus Homo. *Annual Review of Earth and Planetary Sciences* 37:67–92.

Taylor, G.J. 1994. Ursprung und Entwicklung des Mondes. *Spektrum der Wissenschaft.* Sept. 1994:58–65.

———. 2009. Ancient Lunar Crust: Origin, Composition, and Implications. *Elements* 5 (1):17–22.

Taylor, S.R. 1987. The Origin of the Moon. *American Scientist* 75 (5):468–477.

Taylor, S.R., G.J. Taylor and L.A. Taylor. 2006. The Moon: A Taylor Perspective. *Geochimica et Cosmochimica Acta* 70 (24):5904–5918.

Thirlwall, A.P., and Penélope Pacheco-López. 2008. *Trade Liberalisation and the Poverty of Nations.* Cheltenham, UK; Northampton, MA: Edward Elgar.

Thomson, Keith Stewart. 1991. *Living Fossil: The Story of the Coelacanth*. New York: W.W. Norton.

Turekian, Karl K. 1976. *Oceans*. 2d ed, Prentice-Hall foundations of earth science series. Englewood Cliffs, NJ: Prentice-Hall.

———. 1985. *Die Ozeane*: 18 Tab, Geowissen kompakt, Stuttgart: Enke.

Ulex, K. 1989. *Feldmessen. Anleitung zum Feldmesspraktikum der 10. Klasse*. Kassel: Pädagogische Forschungsstelle Abt. Kassel, Bund der Freien Waldorfschulen.

United States Central Intelligence Agency. *The World Factbook*. Central Intelligence Agency.

United States.Energy Information Administration. *Annual energy review*. Energy Information Administration, Office of Energy Markets and End Use. Available from http://purl.access.gpo.gov/GPO/LPS62045.

———. 2007. Annual energy outlook retrospective review evaluation of projections in past editions (1982–2006). [Washington, DC]: Energy Information Administration. http://purl.access.gpo.gov/GPO/LPS89004.

Verband Schweizerischer Elektrizitätswerke. 1988. Strom 2005: *So sehen die zehn grössten Elektrizitätswerke die Stromversorgung der Schweiz bis ins Jahr 2005: Zahlen und Fakten aus dem 7. Zehn-Werke-Bericht des VSE*. Zürich: VSE.

Verhulst, J. 1993. Geologie und Anthroposophie. *Lehrerrundbrief des Bundes der Freien Waldorfschulen*. Okt. 1993:31–42.

Vidal, G. 1984. The Oldest Eukaryotic Cells. *Scientific American* 250 (2):48–57.

Vogel, A. 1991. Convection Tectonics - Global Tectonics and Earthquake Activity in the Light of Mantle-Wide Convection. *Proceedings of the First International Conference on Seismology and Earthquake Engineering, Vols 1 and 2*: 105–124.

Vogel, Andreas. 1993. *Comprehensive approach to earthquake disaster mitigation, Progress in Earthquake Research and Engineering V. 4*. Braunschweig/Wiesbaden: Vieweg.

Volcano. the eruption of Mount St. Helens. 1980. *Volcano, the Eruption of Mount St. Helens*. Longview, WA; Seattle: Longview Pub. Co.; Madrona Publishers.

Volquardts, Hans and Kurt Matthews. 1975. *Vermessungskunde: für die Fachgebiete Hochbau, Bauingenieurwesen, Vermessungswesen*. Stuttgart: Teubner.

Wachsmuth, Guenther. 1950. *Die Entwicklung der Erde: Kosmogonie und Erdgeschichte, ein organisches Werden*. Dornach: Philosophisch-Anthroposophischer Verlag am Goetheanum.

———. 1952. Wie alt ist die Erde? *Beiträge zur Substanzforschung* 1:7–18.

———. 1980. *Erde und Mensch: ihre Bildekräfte, Rhythmen und Lebensprozesse*. [4., durchges. Aufl.] ed. Dornach: Philosophisch-Anthroposophischer Verlag.

Wagner, Günther A. 1998. *Age Determination of Young Rocks and Artifacts: Physical and Chemical Clocks in Quaternary Geology and Archaeology*. Berlin: Springer.

Walker, A. 2002. New perspectives on the hominids of the Turkana Basin, Kenya. *Evolutionary Anthropology* 11:38–41.

Walker, A.C. 1986. Homo-Erectus Skeleton from West Lake Turkana, Kenya. American *Journal of Physical Anthropology* 69 (2):275–275.

Walker, A. and E.F. Leakey. 1978. Hominids of East Turkana. *Scientific American* 239 (2):54–66.

Wanke, H. and G. Dreibus. 1983. The Origin of the Moon. *Fortschritte der Mineralogie* 61 (1):215–216.

Wanke, H., H. Palme, H. Baddenhausen, G. Dreibus, H. Kruse and B. Spettel. 1977. Element Correlations and Bulk Composition of Moon. *Philosophical Transactions of the Royal Society of London Series a: Mathematical Physical and Engineering Sciences* 285 (1327):41–48.

Wegener, Alfred. 1915. *Die Entstehung der Kontinente und Ozeane, Sammlung Vieweg*. Braunschweig: Vieweg.

———. 1970. *The Origin of Continents and Oceans*, University paperbacks. New York; London: Metheun.

Wielen, Roland. 1997. *Planeten und ihre Monde: die großen Körper des Sonnensystems*. 2. Aufl. ed. Heidelberg: Spektrum der Wiss.

Wiggins, S. 2003. Trade reform, agriculture and poverty in developing countries: A review of the empirical evidence. *Agricultural Trade and Poverty Making: Policy Analysis Count*, 287–296.

Witzenmann, Herbert. *Strukturphänomenologie: vorbewusstes Gestaltbilden im erkennenden Wirklichkeitenthüllen: ein neues wissenschaftstheoretisches Konzept im Anschluss an die Erkenntniswissenschaft Rudolf Steiners* (1. Aufl.) G. Spicker 1983.

———. 1987. *Goethes universalästhetischer Impuls: die Vereinigung der platonischen und aristotelischen Geistesströmung. 1.* Aufl. ed. Dornach: G. Spicker.

Zehnder, A.J.B., H. Yang and R. Schertenleib. 2003. Water issues: the need for action at different levels. *Aquatic Sciences* 65 (1):1–20.

Ziegler, Renatus. 1998. *Morphologie von Kristallformen und symmetrischen Polyedern: Kristall- und Polyedergeometrie im Lichte von Symmetrielehre und projektiver Geometrie.* Dornach: Philosophisch-Anthroposophischer Verlag am Goetheanum.

Zürich (Kanton) Amt für Technische Anlagen und Lufthygiene. 1988. *Fakten und Meinungen Ästhetik, Wirtschaftlichkeit und Erntefaktor von Sonnenenergieanlagen.* Zürich: Amt für Technische Anlagen und Lufthygiene des Kantons Zürich.

# Index

## About the Author:

Hans-Ulrich Schmutz, PhD, was born in 1945. He studied Geology at the Swiss Federal Institute of Technology in Zürich, Switzerland, and worked seven years in social work with university students. He taught Geography and Technology in high school for eighteen years at the Rudolf Steiner School in Wetzikon near Zürich. Since 1986, Dr. Schmutz has been involved in teacher training at the Freie Hochschule für Anthroposophische Pädagogik [Free University for Anthroposophic Pedagogy] in Mannheim, the Institute for Waldorf Pedagogy in Witten-Annen, and the teacher seminar in Kassel, all in Germany, and since 1991 at the teacher seminar of the Waldorf Center in Moscow. In addition, he is actively conducting research in the fields of geology and education.

Book publications include: *Die Tetraederstruktur der Erde [The Tetrahedral Structure of the Earth]*, Stuttgart 1986; *Erdkunde in der 9. bis 12. Klasse an Waldorfschulen [Earth Science for Grades 9 to 12 at Waldorf/Steiner Schools]*, Stuttgart 2001; *Acht geologische Exkursionen durch die Alpen [Eight Geologic Field Trips through the Alps]*, Stuttgart 2005.

## About the Translator:

Thomas Wassmer, PhD, was born in 1962 and studied Biology and Cultural Anthropology at the Albert-Ludwigs University in Freiburg and the University of Konstanz, Germany. He worked nine years as a research and teaching assistant and project manager at the Universities of Freiburg, Konstanz, and Mainz and at the Film Archive of Human Ethology of the Max-Planck-Society where he spent two years in the Waldorf teacher training at the Free University for Anthroposophic Pedagogy in Mannheim. He taught high school teacher Biology, Earth Science, and Chemistry at the Rudolf Steiner School of Ann Arbor, MI, from 2006–2010. He is now assistant professor of biology at Siena Heights University, Michigan.

Publications include: "Seasonality of Coprophagous Beetles in the Kaiserstuhl Area Near Freiburg [SW Germany] Including the Winter Months," *Acta Oecologica 1994*, 15: 607–631; "Body Temperature and Above Ground Patterns During Hibernation in European Hamsters [*Cricetus cricetus L.*]." J. Zool., London, 2004, 262:281-288; Die zeitliche Organisation des Winterschlafs beim Europäischen Feldhamster *Cricetus cricetus L.* ["The Temporal Organization of Hibernation in the European Hamster *Cricetus cricetus L*]," München 1998.

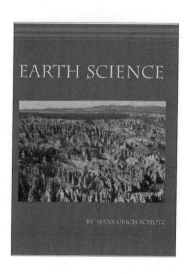

**About This Book:**

Hans-Ulrich Schmutz has developed a comprehensive curriculum for teaching earth science in Waldorf/Steiner high schools which is entirely consistent with the developmental stages of high school students. His consistent and carefully developed curriculum guides the student from geology through the dynamics of ocean currents and global climatology, crystallography, technology, and economic geography to astronomy and paleontology. All these subjects are provided with new and interesting point of views, which gradually contribute to a discovery of the living and the connection between earth and the human being.

Made in the USA
Lexington, KY
09 December 2019